T0327365

Register-based Statistics

Register-based Statistics

Administrative Data for Statistical Purposes

Anders and Britt Wallgren

Statistics Sweden, Sweden

John Wiley & Sons, Ltd

Other Wiley Editorial Offices

John Wiley & Sons Inc., 111 River Street, Hoboken, NJ 07030, USA

Jossey-Bass, 989 Market Street, San Francisco, CA 94103-1741, USA

Wiley-VCH Verlag GmbH, Boschstr. 12, D-69469 Weinheim, Germany

John Wiley & Sons Australia Ltd, 42 McDougall Street, Milton, Queensland 4064, Australia

John Wiley & Sons (Asia) Pte Ltd, 2 Clementi Loop #02-01, Jin Xing Distripark, Singapore 129809

John Wiley & Sons Canada Ltd, 6045 Freemont Blvd, Mississauga, ONT, L5R 4J3

Wiley also publishes its books in a variety of electronic formats. Some content that appears in print
may not be available in electronic books.

British Library Cataloguing in Publication Data

A catalogue record for this book is available from the British Library

ISBN-13 978-0-470-02778-3 (HB)
ISBN-10 0-470-02778-9 (HB)

Typeset in 10/12pt TimesNewRoman by the authors.

Contents

Preface

Register-based surveys are becoming more and more common within a growing number of national statistical offices, but are also common within enterprises and other organizations, where data from the organization's own administrative systems are used to produce statistics on, for example, production, sales and wages.

Although register-based statistics are the most common form of statistics, no well-established theory in the field has existed up to now. There have been no well-known terms or principles, which have made the development of both register-based statistics and register-statistical methodology all the more difficult. As a consequence of this, ad hoc methods have been used instead of methods based on a generally accepted theory.

Many countries are investigating the possibilities to use more and more administrative data for statistical purposes. It is necessary to reduce response burden and costs; increasing nonresponse in censuses and sample surveys also make this new strategy necessary. A new approach is necessary and register-based surveys require that suitable statistical methods be developed.

We have studied the requirements for register-based statistics through analysis of Statistics Sweden's system of statistical registers. During more that ten years, we have devoted an increasing part of our work, at the Department of Research and Development at Statistics Sweden, to the study of register-based surveys. We have also worked together with a number of manufacturing enterprises and analysed their administrative data for the purposes of their management. These experiences are also used in this book.

One purpose of the book is to describe the register system and discuss how it should work, presenting the possibilities offered by a functioning register system. Another purpose is to structure and describe the register-statistical methodological work which will provide the basis for the creation of the system. In several cases, we suggest new terminology. The book can be considered as a first step to a more systematic way of working with register-statistical issues. Our hope is that it will stimulate further development in this important area and encourage an overview of and improvements to the current ways of working. The necessary further development, among other things to ensure a better consistency between the different register-based surveys, is a task that may take many years.

Professor Carl-Erik Särndal has been a very important discussion partner during our work with the book. His broad experience from statistical offices in different countries and his background as a specialist in sample surveys have been enormously useful. In addition, around 50 persons within Statistics Sweden have read and commented on different parts of the first Swedish version of this book. Several individuals have also been interviewed to provide material for different examples and methodological sections.

Register-based Statistics – Administrative Data for Statistical Purposes   A. Wallgren and B. Wallgren
© 2007 John Wiley & Sons, Ltd

The Swedish book was published 2004, and has been used in a number of study circles within Statistics Sweden. These study circles were very stimulation and have helped us in our work with this English version.

When preparing this version of the book we have used many valuable comments and suggestions from five anonymous reviewers engaged by John Wiley & Sons, Ltd.

Section 11.6, *IT Systems for Register-based Statistics*, is written by Lars-Göran Lundell, Statistics Sweden, and we have discussed many parts in the book with him to get the IT perspective on the register system.

The comments and encouragement from all these people mentioned above have been of great help in the work with the book.

It is our hope that *Register Statistics – Administrative Data for Statistical Purposes* and its proposals will stimulate the discussion on register statistics.

Örebro, Sweden   Anders Wallgren
January 2007   Britt Wallgren

CHAPTER 1

Register-based Surveys – An Introduction

This chapter and the next introduce a variety of concepts and principles that will be used in this book when discussing *register-based surveys,* that is, surveys that are based on data from administrative registers. These concepts and principles form the basis for a theory on this type of survey.

Register-based surveys are common within enterprises and other organisations, where data from the organisation's own administrative systems are used to produce statistics on for instance production and sales. Register-based surveys are also common at the national statistical offices in the Scandinavian countries, where many administrative registers are used to produce official statistics.

In this book, we will primary discuss register-based surveys at national statistical offices. There is an increasing interest in this area; many countries use more and more administrative data for statistical purposes and there is a growing demand for a theory on register-based surveys.

Our aim is to present statistical methods and principles of general interest, but we will use Scandinavian experiences and case studies from Statistics Sweden to illustrate these general methodological issues.

1.1   DO WE NEED A THEORY ON REGISTER-BASED SURVEYS?

Within the national statistical offices, three kinds of statistics are published – statistics based on sample surveys, statistics based on censuses and statistics based on administrative registers. It is most common to only differentiate between sample surveys and censuses, where the statistical office is responsible for the collection of the data. These two survey types are dominated by the work to collect data.

However, this book deals with the third type of statistics that are based on administrative registers where, instead of collecting data through surveys and censuses, administrative registers from different sources are adapted and processed to be suitable for statistical purposes. This kind of survey is called a *register-based survey*.

Sample surveys are based on methods that have been derived from an established theory – *sampling theory*. This theory has been developed both within the academic world and within statistical offices, and consists of terms and principles that are generally well known.

Register-based Statistics – Administrative Data for Statistical Purposes   A. Wallgren and B. Wallgren
© 2007 John Wiley & Sons, Ltd

Scientific literature and journals develop and spread the methodologies for sampling and estimation. Because the terms and principles are well known, persons working with sample surveys can easily communicate and exchange their experiences.

Censuses with their own data collection are based on a long tradition of population censuses and the collection of data from local authorities, schools and different types of enterprises. Measurement errors, design of questionnaires and nonresponse are methodology issues that also apply for sample surveys. Censuses and sample surveys are closely related in terms of methodology – censuses are often considered as special cases where the sample is the entire population.

Statistics based on administrative registers will hereafter be called *register-based statistics*. Although this is the oldest and most common form of statistics, no well-established theory in the field exists. There are no well-known terms or principles, which makes the development of both register-based statistics and register-statistical methodology all the more difficult. As a consequence of this, *ad hoc methods are used instead of methods based on a generally accepted theory.*

One important reason for this shortfall is that the subject field of register-based surveys is not included in academic statistics. Statistical theory within statistical science is understood to consist of *probability theory* and *statistical inference.* Sampling theory is included within this theoretical school of thought, but register-based surveys based on total enumeration are not.

Unfortunately statistical science has so far not included any theory on statistical systems. Statistical offices, larger enterprises and organisations do not carry out separate surveys so often. It is more common that statistical information systems are built, which constantly generate new data. A statistical theory is necessary to describe the general principles and to develop the concept apparatus for such statistical systems. Register-based surveys should be included in this theory.

In 1995, Statistics Denmark published *"Statistics on Persons in Denmark – A Register-based Statistical System".* The Danish book presents a systematic review of register-statistical work and describes how to design a well-prepared register system.

In this book, we build on and add to the Danish work. The next chapters introduce a number of register-statistical concepts and principles. The Glossary compiles all these concepts and terms. The aim is that all those working with the development of register-based surveys could then use the terms generally.

We formulate four principles for how administrative registers should be used:

Chart 1.1 Four principles on how to use administrative data

1. A statistical office should have access to administrative registers kept by public authorities. This right should be supported by law as the protection of privacy.

2. These administrative registers should be transformed into statistical registers. Many sources should be used and compared during this transformation.

3. All statistical registers should be included in a coordinated register system. This system will ensure that all data can be integrated and used effectively.

4. Consistency regarding populations and variables are necessary for the coherence of estimates from different register-based surveys.

We will use these principles in the book and gradually introduce the register-statistical terms that are needed for the discussions.

1.2   WHAT IS A STATISTICAL SURVEY?

The starting point for any survey is a number of questions in connection to a specific area of interest. A survey is carried out to try and answer these questions. The survey process can be described in more or less detail. Simply described, the work consists of the following phases:

1. Determining the research objectives and planning of the survey.
2. Procurement and processing of data.
3. Estimation, analysis of data and presentation of the results.

Within a national statistical office it is usual to work with surveys, which are repeated every year, quarter or month. With such surveys, work is mainly carried out in phases 2 and 3. However, these surveys have also had a phase of determining objectives and planning, even if this was a long time ago.

A separate survey can be a commission where the statistical office is to carry out the entire survey and this involves working with all the three phases. However, in many commissions, it is the customer who carries out phases 1 and 3 and the statistical office is only brought in to work with phase 2.

Phase 2 of a survey, the procurement of data, can be carried out in different ways:

a. With own data collection using a *sample survey*.
   Example: The Labour Force Survey is conducted in many countries. A new sample is taken monthly, with new data collection and reporting.
b. With own data collection using a *census*.
   Example: The traditional Population and Housing Census, in which all households and house owners are interviewed or asked to complete a questionnaire which is then processed by the national statistical office.
   Because censuses result in the creation of a register, microdata from censuses are also included in the system of statistical registers and can therefore form the basis for register-based surveys.
c. Existing microdata is used for a *register-based survey*.
   Microdata refer to data on individual *objects*. Existing administrative or statistical registers with data that, for example, refer to individual persons or enterprises are used for the purposes of the register-based survey.
   Example: In Section 1.4 below we give two examples of how statistical registers are created to meet the needs of different register-based surveys.

Because these three types of surveys differ in terms of methodology, it is appropriate to differentiate them conceptually. Sample surveys, censuses and register-based surveys are the most important types of surveys at a national statistical office.

A statistical population consists of *N objects* or *units* or *elements*. Of these three synonyms we will as a rule use the term *object* in this book.

1.3   WHAT IS A REGISTER?

An *administrative register* is maintained to store records on *all* objects to be administered and the administrative process requires that it is possible to *identify* all objects. The following definition is valid for both administrative and statistical registers:

> A *register* aims to be a complete list of the objects in a specific group of objects or population. However, data on some objects can be missing due to quality deficiencies. Data on an object's identity should be available so that the register can be updated and expanded with new variable values for each object. Complete listing and known identities are thus the important characteristics of a register.

The identities used in register processing can either be identity numbers who are unique within a national administrative system or an identity number in a subsystem with keys to the identities in other systems. It is also possible to use identities defined by for instance name, address, date of birth and birthplace.

These identities will be used in exact matching of the objects in different registers, where the aim is to find identical or related objects in two registers.

A *statistical register* is based on data from administrative registers that have been processed to suit statistical purposes. The register processing which transforms administrative data into statistical registers gives rise to important methodological questions that will be discussed later in the book.

The term *statistical register* is used to describe registers within a system of statistical registers within a statistical office or other organisation. Such registers can be based either on a census carried out by the agency or on administrative registers from authorities and organisations outside the statistical office.

Data collection in a sample survey does not give rise to a register, as the micro data about the sample only consists of a small part of the surveyed population. Chart 1.2 compares the three types of survey that dominate at national statistical offices.

Chart 1.2 Comparison between the three types of survey

Sample survey	Census	Register-based survey
Not in the register system	Included in the register system	
Own data collection		Uses existing registers

The term *register-based statistics* refers to statistics that are based on register-based surveys. When we discuss the register system, as in Chapter 2, we do not differentiate between censuses and register-based surveys. However, when we discuss methodology issues, the term only refers to register-based surveys.

1.4   WHAT IS A REGISTER-BASED SURVEY?

Administrative registers are created and delivered to a national statistical office
The original data formation is carried out in the authorities and organisations. The definitions of objects and variables are adapted to administrative purposes. Every authority

carries out controls, corrections and other processing that are suited to their administrative aims. When an authority delivers data to a national statistical office, further selections and processing may be carried out to meet the needs of the statistical office. The respective authorities also have metadata in the form of information on the definitions, data formation and quality. This type of information is also important for those receiving the data within the statistical office.

What happens when data is delivered to a statistical office as Statistics Sweden?
It is generally not a good idea to produce statistics directly from the received administrative registers because these are not adapted to statistical requirements. The object sets, object definitions and variables need to be edited and it will often be necessary to carry out some processing so that the register fulfils the statistical requirements for objects and variables. The register-statistical processing, which aims to transform one or several administrative registers into one statistical register, should be based on generally accepted *register-statistical methodology*. These methodological issues are discussed in more detail in the following chapters. The chart below shows the different elements included in statistical methodology work.

Chart 1.3 From an administrative register to a statistical register

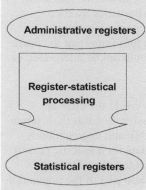

	Register-statistical processing:	Quality assurance:
Administrative registers	The administrative registers are processed so that objects and variables meet statistical needs: · Editing of data · Coding of variables · Handling of missing objects and missing values · Matching and selections · Processing of time references · Creating derived objects · Creating derived variables The statistical register is used to produce statistics	· Contacts with data suppliers · Checking received data · Missing values: reasons and extent · Causes and extent of mismatches · Evaluate quality of objects and variables · Register maintenance surveys · Inconsistencies are investigated and reported

In the next two subsections, we describe how two statistical registers are created. The examples are from Statistics Sweden, but they illustrate general principles. Each of these registers is created to meet the needs of a number of register-based surveys. The examples illustrate how administrative data are transformed to meet statistical needs and how the system of statistical registers (at Statistics Sweden) is used when creating statistical registers. The main part of the work with a register-based survey is the work spent on creating an appropriate register.

1.4.1 Statistics Sweden's Income and Taxation Register

This register utilises many administrative sources. Many administrative variables are used to create important statistical variables. Besides these administrative sources it is necessary to use the register system at Statistics Sweden: the Population Register is used to define the population of the Income and Taxation Register, and important classification variables are imported from other registers in the system to the Income and Taxation Register.

1. *Data formation at the National Tax Board*
    The annual income assessment is based on tax declarations from income earners

and the taxation decisions of the local tax authority. Both the income earner and the tax authority use statements of earnings regarding salary, sickness benefit and interest that the employers, social insurance office and finance companies are responsible for. The National Tax Board ultimately compiles this information. Declarations, statements of earnings and taxation decisions can be changed and supplemented. Data for one person can thus be very complex.

2. *Microdata deliveries to the Income and Taxation Register*
The Swedish National Tax Board annually creates databases that contain information on Sweden's population. The data files for one year – containing around nine million records, each with around 300 variables – are delivered directly to the Income and Taxation Register at Statistics Sweden.

3. *Metadata to the Income and Taxation Register (I&T)*
Record descriptions with variable names and variable definitions accompany the deliveries from the National Tax Board. Tax declaration forms, statement of earnings forms, taxation decisions, tax declaration instructions and instructions to employers are also needed to be able to interpret the data.

4. *Editing of data*
The I&T Register receives data from eleven different suppliers both outside and inside Statistics Sweden. Data from outside is edited. Data from other Statistics Sweden registers has already been edited. Contacts with suppliers are important to obtain knowledge of changes in the administrative system, which is in turn important to ensure the quality of the register statistics – administrative changes should not be interpreted as actual income changes.

5. *Matching and selections*
There is a large number of registers that should be processed to create the different sub-registers that are included in the Income and Taxation Register. Records in different sources are matched using Personal Identification Numbers (PIN), and aggregation is carried out at the same time, i.e. all the statements of earnings data for a specific person are aggregated so that the person's income from work can be put together. One type of processing is to select persons aged 16 and older, who were also part of the population on December 31.

6. *Derived objects are created*
More information on certain relations helps to form household units. Between adults, the relations *married* or *cohabiting adults with children in common* result in that they are placed in the same household unit. These relations are shown by the family members' personal identification numbers, these reference variables are found in the taxation data and in Statistics Sweden's Population Register.

7. *Derived variables are created*
A large number of derived income variables are formed. For instance, the wage or salary amounts are aggregated from the different earnings data to become an individual's *income from work*. Every person's total income from work and capital plus transfer payments minus tax becomes the person's *disposable income*. For households, variables such as *household type, number of consumption units* and *disposable income* are formed.

The chart below shows how the Income and Taxation Register receives administrative data from a variety of different external sources and some Statistics Sweden registers. The term *source register* refers to the administrative sources and the Statistics Sweden registers that are used to create the new register. The different phases when the source registers are used during the process to create the new statistical register are shown in the middle column in the chart.

This example shows the importance of the four principles in Chart 1.1. Statistics Sweden has access to many administrative registers with variables describing different kinds of income. The object set and the administrative variables have been processed to meet statistical needs. Many sources have been used to produce a statistical income register with rich content. The population in the income register is consistent with other statistical registers within the register system.

Chart 1.4 Different data sources for the Income and Taxation Register (I&T)

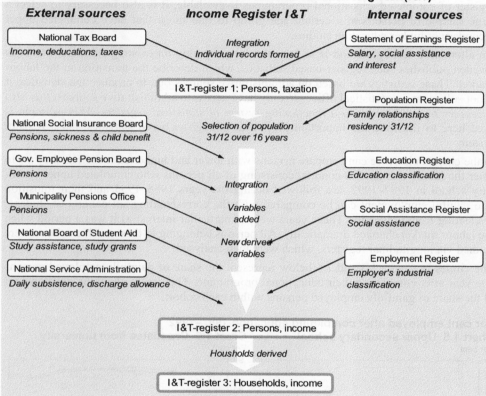

The Income and Taxation Register is an important part of Statistics Sweden's register system. It is used to describe the income distribution, for regional income statistics and it is also the basis for longitudinal income registers used by university researchers.

1.4.2   Longitudinal register – education and labour market

The Income and Taxation Register mentioned above is directly based on large amounts of administrative data. However, many important registers at Statistics Sweden are not directly based on administrative data; they are instead based on already existing statistical registers in the register system. The example illustrates how existing data can be used in a new and more advanced way after specially adapted register processing.

The entry of young persons into the labour market after completing their studies is nowadays an important area for different surveys. Such surveys should be carried out as *longitudinal surveys*, where groups of persons are followed over a period of years. If these surveys are carried out as sample surveys, a sample is taken every year among persons completing a specific educational programme and each sample is interviewed or asked to fill in a questionnaire once a year for a period of years, in this case seven years.

This survey method has its disadvantages, partly that the burden on the respondents is heavy – the selected individuals must answer a large number of questions every year – and partly that nonresponse will gradually increase over the period. In addition, if no adequate register of completed educational programmes is available, it is also necessary to select a large sample of persons in a certain age group to find those that have completed upper secondary or higher education studies.

An alternative survey method is to base the survey purely on existing registers. Statistics Sweden publishes such register-based statistics, which describe the transition to the labour market. These statistics are based on administrative sources but, to produce the statistics, it is not sufficient for statistical offices to only have access to administrative sources. *It is also necessary to have access to a functioning system of statistical registers.* This example is used here to illustrate the important properties of register-based statistics and a register system.

In the charts below, we can compare persons with lower and higher education as they try to enter the labour market. Six cohorts, consisting of all persons who completed upper secondary school in 1987–1992, are followed during the years 1988–1993 and their transition into gainful employment can be compared with the corresponding six cohorts of students graduating from university. These years were of particular interest as it was a period when the labour market changed dramatically. All persons belonging to these twelve cohorts were studied via longitudinal registers, which were then analysed.

The circles in Charts 1.5 and 1.6 below represent the share of gainfully employed persons *one* year after completing their educational programme. The curves show the development of the share of gainfully employed persons within each cohort.

Per cent employed after completing education 1987–1992

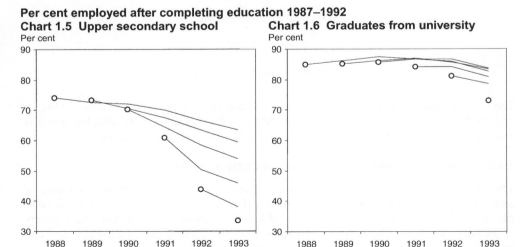

Chart 1.5 Upper secondary school   **Chart 1.6 Graduates from university**

At the beginning of the 1990s, the most serious crisis in the Swedish labour market since the 1930s occurred. Charts 1.5 and 1.6 show how the economic downturn at the beginning of the 1990s changed the possibilities for young persons to enter the labour market. The share with gainful employment one year after completing upper secondary studies changed dramatically during these years. For those with university degrees, the development was not so depressing – this appears to show that higher education studies were worthwhile.

This example is based on longitudinal data. Every year, each person is classified as gainfully employed, studying in higher education or neither gainfully employed nor studying. For gainfully employed persons, the annual income from work is registered as well as information on which sector the person is working in. Persons are also classified by course/study programme, sex and region, which permit detailed reporting. The annual cost for this statistical product was SEK 0.4 million (approximately $ 50 000), half of which was printing cost.

How was this register-based survey carried out?
The charts above show the transition of young persons from education to the labour market. The longitudinal register that the charts are based on was created in the following way:

– By combining information from three Statistics Sweden registers for 1987–1993 (i.e. a total of 21 different registers), a new *integrated register* was created, which is marked in a bold frame in Chart 1.7 below.

– The objects in the new register were created by selecting certain objects in the Register of Education.

– Variable values are imported into the new register by *matching* the objects in the new register with the corresponding objects in the Education Register, the Register of University Students and the Employment Register. This is illustrated in the chart below.

Chart 1.7 Creating an integrated register for the transition to the labour market

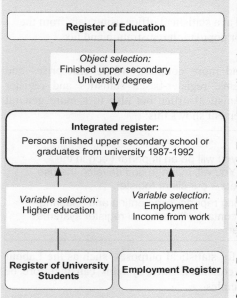

Register of Education
Based on data from schools or universities
Object: Person
Variables: Personal identification number, education, year of degree
Processing: Selection of persons who finished upper secondary school or graduated from University

Register of University Students
Based on data from universities
Object: Person actively studying
Variables: Personal identification number, pursuing higher education
Processing: Matching against selected group. Variables are selected and transferred to the integrated register

Employment Register
Based on statements of earnings, income, population and business registers
Object: Person
Variables: Personal identification number, employment status, income from work
Processing: Matching against selected group. Variables are selected and transferred to the integrated register

We have mentioned here a variety of important terms. By bringing together or integrating information from several registers, an *integrated register* is created. We distinguish between *object selection* where certain objects are selected and *variable selection* where certain variables are selected. *Matching* means that the links in two or more registers are compared, resulting in a match or mismatch.

To create the new integrated register, no new administrative sources were used. The register was instead based on the increased usage of existing registers. The Employment Register is also based on existing registers within Statistics Sweden's register system. The Register of University Students, however, is built from administrative data that Statistics Sweden receives from authorities outside Statistics Sweden.

To create and coordinate these registers, extensive processing of register data in several registers is necessary. This means that having access to administrative sources is not sufficient to be able to produce these statistics. Without Statistics Sweden's coordinated system of statistical registers, it would be impossible to produce these statistics. When the register system is used, exact matching of records in different registers is done with standardised identification numbers. In the example above, Personal Identification Numbers (PIN) are used.

1.5   ADMINISTRATIVE AND STATISTICAL INFORMATION SYSTEMS

Using administrative data for statistical purposes is not something specific to Statistics Sweden or to statistical offices in other countries. It is also a common practice in large enterprises and organisations. Administrative systems are used generally as sources for statistical information and there is no major difference between the following enterprise example and register-based statistics at a national statistical office:

– Statistics on staff and salaries within an enterprise can be produced using the personnel management system.
– Population and income statistics are produced at a statistical office using data from the National Tax Board's tax collection system for population registration and tax assessment.

Register-based surveys have become more and more common within enterprises and organisations. Knowledge about register systems, register-based statistics and register quality is not only needed within a national statistical office but also in a more general sense. The following extract from a job advertisement shows this:

> **Market analyst**
>
> As an analyst in the marketing department, you will be an important cog in the wheel of our enterprise's continued growth. You will manage and develop the use of one of the enterprise's most valuable assets – our client register.
>
> You will work with campaign analyses, drafting reports, segmenting and ensuring the quality of the register. You will maintain contact with external register systems and work closely with the marketing manager.

Certain information systems are built up solely for statistical purposes, such as the Labour Force Surveys, which are conducted in many countries. Such systems can therefore be completely designed according to statistical principles.

Other information systems are used both for administrative and statistical purposes, which can sometimes lead to conflicts with regard to the structure of the system. In general, these systems are primarily intended for administrative purposes and the statistical information is a by-product.

However, there are several differences between a pure administrative system and a pure statistical system. These two kinds of systems are compared below.

1.5.1 Different purposes

Information in an administrative system is used as a basis when taking administrative measures and decisions that will affect the objects in the system.

Example: A personnel management system is used to carry out salary payments.

Information in a statistical system is used as the basis for analysis and from the analysis one will draw conclusions. These conclusions can then become the foundation for policy-related decisions.

Example: A statistical salary system is used to study salary structure. How has this changed? What are the differences in monthly salaries between different categories of staff? This analysis could then involve a change in policy relating to salary issues, e.g. females should be paid better.

1.5.2 Different roles for individual objects

In an administrative system, decisions are made and measures are taken with regard to individual objects. To this end, information relating to that specific object is retrieved.

Example: Salaries are paid to every employee in an enterprise. Administrative information is checked and the salaries and taxes for the employees can be calculated.

In a statistical system, the individual objects are not interesting in themselves. In a statistical analysis aggregated estimates are calculated and compared for groups of objects.

Example: Salary totals, average salaries, the dispersion of salaries, etc. are calculated for the different categories of staff.

1.5.3 Approaches regarding errors

From an administrative point-of-view, certain items of the individual information must be absolutely correct, but other items can be more approximate. From a statistical point-of-view, errors can exist but they should be carefully controlled and attempts are made to reduce the errors, which can be significant for the statistical conclusions. Errors can be accepted in some data, but only if these are considered to have a limited effect.

Example: The personal identification number in a personnel management system must be completely correct from the point-of-view of salaries and tax administration. As the Swedish personal identification number contains the date of birth, it can also be used to describe the age structure of the staff. If, for example, 30% of the staff has an incorrect number for the month in their registration number, this would not affect the statistical analysis particularly although the salary and tax routines would become impossible.

1.5.4 How should administrative data be processed for statistical purposes?

A general principle is to combine many sources when a statistical register is created. There are many reasons for this: Variables from different sources can be used to achieve a rich content, as in Sections 1.4.1 and 1.4.2; other reasons are that coverage and editing possibilities can be improved.

When the object set in an administrative source is not suitable as a relevant statistical population, a number of sources should be combined to create an object set with good coverage.

Example: A business register at a national statistical office is based on administrative sources. With five sources we created a Business Register for Sweden containing all enterprises (legal units) active during 2002. Each source consists of the legal units in one taxation system. In the table below, undercoverage and overcoverage of the sources are compared with our final Business Register. Source 1 has the earliest and source 5 has the latest available information. The administrative object sets in each source is adequate for each of the five taxation systems. But taken alone, each source is of low *statistical* quality, however, if all sources are combined, the coverage is good.

Chart 1.8 Overcoverage and undercoverage in five administrative sources

	Source 1	Source 2	Source 3	Source 4	Source 5
Overcoverage	41%	0%	0%	0%	0%
Undercoverage	21%	74%	74%	30%	9%

In sample surveys and censuses, editing as a rule uses the collected data only. In register-based surveys, however, it is possible to compare variables from different sources, and this gives better possibilities to find and correct errors. Measurement errors and other kinds of inconsistencies can be detected if different sources are compared.

Example: In Chart 1.9 the enterprise units (each consists of a number of legal units) in a Swedish business survey are compared with respect to turnover reported in two administrative sources.

Due to different definitions, source 2 should report larger turnover than source 1. This means that the four enterprise units above the diagonal line should be checked – probably these enterprise units are incomplete.

Also the enterprise units with large deviations under the diagonal line must be checked – probably the reporting in source 1 has been delayed for some legal units belonging to these enterprise units.

Chart 1.9 Comparing sources

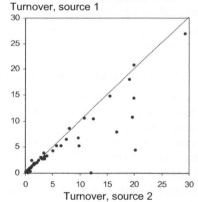

Turnover, source 1

Turnover, source 2

1.6   WHY USE ADMINISTRATIVE DATA FOR STATISTICS?

There are both prejudices and legitimate criticism against register-based statistics. The prejudices consider statistics based on administrative data to be cheap but of bad quality, compared to 'true' survey statistics. The legitimate criticism can refer to relevance errors, comparability problems and that the statistical quality is not under control. Unfortunately, similar criticism can be directed towards sample surveys for which frame errors, measurement errors and nonresponse errors can be significant and undetected, irrespective of whether the sampling errors are under control.

Our answer to these types of comments is that statistics on society should consist of both register-based statistics and statistics based on data collected by a national statistical office. It is therefore not a question of which method is better than the other, but more that in certain situations register-based surveys are more effective and in others, sample surveys are most appropriate.

1.6.1   Pros and cons of register-based surveys

As we have mentioned, there is a common but often diffuse idea that statistics based on administrative data is of low quality. Is this idea justified for the administrative sources that a statistical office uses? A very large part of Statistics Sweden's register system is based on data from the administrative population register and tax administration. Would these statistics be of higher quality if Statistics Sweden collected the data itself in parallel with the National Tax Board's collection of administrative data? This is hardly the case – Statistics Sweden's own attempt to collect these data would be expensive, would increase the burden on the respondents and would likely produce data with more measurement errors.

Chart 1.10 Pros and cons of surveys based on data collection or registers

	Advantages	Disadvantages
Surveys based on data collection: sample surveys and censuses	Can choose which questions to ask Can be up-to-date	Some respondents .. ... do not understand the question ... have forgotten how it was ... do not respond (nonresponse) ... respond carelessly Burden on respondents can be high Expensive Low quality for estimates for small study domains (for sample surveys)
Register-based surveys	No further burden on the respondent for the statistics Low costs Almost complete coverage of population Complete coverage of time Respondents answer carefully to important administrative questions Good possibilities for reporting for small areas, regional statistics and longitudinal studies	Cannot ask questions Dependent on the administrative system's population, object and variable definitions The reporting of administrative data can be slow; the time between the reference period and when data are available for statistical purposes can be long Changes in the administrative systems make comparisons difficult Variables that are less important for administrative work can be of lower quality

In this table, we have summarised the advantages and disadvantages of the two survey methods. The significance of the disadvantages can vary in different survey situations.

Our example of the transition of young persons from education to employment can illustrate both the advantages and disadvantages of register-based statistics. We have already mentioned that register-based surveys are appropriate for longitudinal studies. Another advantage is that it is possible to report results for many sub-groups, different courses, study programmes and regions. But a disadvantage with certain types of administrative sources is that administrative systems often require a long time from the reference period to when the data are available, which can cause delays with register-based statistics. In our example, the report describing the period 1987–1993 was published in June 1995, i.e. one and a half years after the end of the period.

So it is not just a question of *if* administrative data is to be used but also *how* it is to be used. Our response to *how* is that administrative data should in general not be used as it is, but should be processed for statistical purposes. An attempt should be made to build a system of coordinated registers – this can be advantageous both in terms of quality and cost, and quality assurance should be an important component of the system.

There are many development trends that threaten the quality of statistics based on sample surveys or censuses. The increased usage of telemarketing and number presentation on telephones means it is harder to carry out telephone interviews. If respondent motivation decreases, nonresponse and problems with measurement errors will increase. It will become harder to motivate the double provision of data – why respond to a questionnaire on the enterprise's turnover when you also submit a Value-Added Tax (VAT) declaration to the Tax Board which includes the same information? All these circumstances point to an increase in the significance of register-based statistics. Evidence that double provision of data to Statistics Sweden and to another authority is regarded as unreasonable can be seen in this local newspaper clipping:

Mariestads Tidning, 26 June 2000   (*translated from newspaper article*):

Refuse to send statistics to Statistics Sweden!

Mr R from the B-farm thinks that the authorities should be able to find the information from their own registers.

Mr R refuses to send in statistics to Statistics Sweden. Because he already sends in information every other week to the Swedish Board of Agriculture, he thinks that the authorities should cooperate with each other instead. ...

1.6.2   The cost aspect – are register-based statistics cheap?

"It is quite clear that the Member States find themselves in the paradoxical situation of having to face a number of budget cutbacks at the same time as providing users with an increasing volume of high-quality relevant information." So began Yves Franchet, then Director General for Eurostat, a seminar (Eurostat, 1997) on the use of administrative sources for statistical purposes. The quote illustrates the need for more effective statistical systems. These requirements for increased efficiency can be met by combining two strategies:

– By using more administrative sources, the cost of data collection is reduced both for the respondents and the statistics producers.

– A more effective and flexible usage of existing data would mean that new requests could be met without the costly collection of new data. This could be achieved by using the administrative sources to create a system of coordinated statistical registers.

The construction of such a register system would be costly but, with increased use of the system, the marginal costs would decrease. In the example above of the transition of young persons from education to working life, the marginal cost of the product was small.

The burden on respondents is also a cost that can be reduced by moving from statistics based on data collection to register-based statistics.

Sample surveys are an increasingly expensive method; the number of telephone calls to first contact increase and the costs of reminding an increasing percentage of respondents who don't send back questionnaires are also rising.

The costs for sample surveys per inhabitant are higher for smaller countries – for corresponding domains of study; the same level of accuracy requires almost the same sample size in a small country as in a larger country. Therefore, it becomes especially important for smaller countries to build up a statistical register system.

1.7   AN OVERVIEW OF THIS BOOK

Every chapter in this book contains proposals for change. A new approach, new terminology and new methodologies are needed so that the register system and register-based statistics can be developed and function in an even better way than they do today. Below we provide a summary and overview of the contents of the chapters in the book.

Chapter 1   Register-based surveys – an introduction
So that register statistics can be developed, a *register statistical theory* is needed, with statistical systems at the centre. A well functioning register system forms the basis for the effective production of statistics.

Chapter 2   How to structure a register system
The register system's structure is presented here. The *register model* is an important tool to help spread understanding of the register system. The significance of *administrative sources* is discussed and the role of the four *base registers* is defined.

Chapter 3   A terminology for register-based surveys
A number of *register-statistical terms* are introduced; without good terminology, register theory would be vague and the exchange of experiences difficult. Terms for different kinds of registers are presented and variables are defined by their role in the register system. Variables derived via matching are also described.

Chapter 4   Sample surveys and registers
How can sample surveys benefit from the register system and how can sample surveys and registers be combined? An outline of these topics is given and also the differences between the methods used in sample surveys and register-based surveys are discussed. It is important to distinguish between surveys with different methodologies – on the one hand, sample surveys and censuses and, on the other hand, register-based surveys.

The data collection phase in surveys with their own data collection corresponds to the different types of register processing carried out to create a statistical register. This register processing should be studied from a statistical science point-of-view.

Chapter 5   How to create a register – the population
The procedure of creating a register is structured into five phases. For register-based statistics the term *register population* plays in important role, corresponding to the term *frame population* in sample surveys. All available sources should be used to create a register population with good coverage. By combining different administrative sources it is also possible to define a statistically relevant register population. A number of principles are presented, such as *'everyone should use the standardised populations from the base registers'* and *'everyone should support the base registers so that these have as high a level of quality as possible'.*

Chapter 6   How to create a register – the variables
Derived variables are central for register-based statistics. Administrative variables are used to define statistically important derived variables. Editing of administrative data is discussed via a number of case studies. There are some important differences between editing of data in a sample survey and editing of administrative data – consistency editing of many sources at the same time, and searching not only for variable errors but also for object errors.

Chapter 7   Estimation methods
Estimation methods are also necessary within register-based statistics. Simply summing the data is not always sufficient. The traditional approach within register statistics is that no estimation methods are needed, but the same statistical approach that are currently used by those working with estimation problems for sample surveys and censuses should also be applied to register statistics.

We differentiate between basic and supplementary estimation methods. The supplementary methods use weights in a similar way as weights are used in sample surveys. How weights are used and the calibration of weights is illustrated by examples.

Chapter 8   Calibration and imputation
Different estimation methods can be developed to deal with problems due to nonresponse or *missing values, overcoverage,* and *level shifts in* time series.

When different registers are integrated and variables are imported from one register to other registers, quality flaws such as missing values are also imported to these other registers. This means that it is not sufficient to adjust for missing values in a variable only in one register, the adjustment method must adjust for missing values in this variable in the whole register system in a consistent manner.

Overcoverage can cause serious errors in register-based statistics. We propose that calibration of weights can be used to adjust for overcoverage in the base register where the problem exists. All other statistical products using the base register will then use these weights.

Calibration of weights can also be used to adjust for level shifts in time series at the micro level. With these weights, consistent and linked time series can be produced.

Chapter 9   Estimation with combination objects
Multi-valued variables, such as Industry, are today used in a way that gives rise to aggregation errors. Special estimation methods using weights are introduced in this chapter that can be used for multi-valued variables to reduce these aggregation errors.

Combination objects are introduced for the estimation issues that are related to multi-valued variables. Such combination objects can also be used to adjust for level shifts in time series at the micro level.

Chapter 10   Quality in register-based statistics
Sampling errors have for a long time been regarded as the most important error in sample surveys. In register-based surveys there is no sampling phase; instead this kind of survey is dominated by the *integration phase*, where data from different sources are integrated into a new statistical register.

During the integration phase the register population and derived objects are created, variables are imported from different sources and derived variables are created. The kinds of errors that have their origin from the integration phase should be called *integration errors*.

Specific *quality indicators* are needed that suit the needs of register-based statistics. A number of indicators are presented and, for a specific register, the important indicators for that particular register are selected. Additionally, an *overall assessment* of quality should be made.

Chapter 11   Metadata and IT-systems

Register statistics requires a metadata system in which large amounts of *formalised metadata* can be processed using the appropriate IT tools. A *calendar* with the important changes and a *definitions database* are important parts of such a system.

The IT system for register statistics should be documented in a different way to the traditional *systems documentation*, which relates to surveys with their own data collection. *Data Warehouse technology* can be a tool for:

– more effective register management,

– an increased amount and more uniform metadata,

– simpler and more secure retrievals, and

– a better overview of the system's content.

Chapter 12   Protection of privacy and confidentiality

Well functioning routines for *the protection of privacy* are a very important part of the register system. Minimising the use of variables with information in plain language and official identification numbers should be considered. Routines for the *protection of disclosure* should always be included in the work to present and disseminate tables based on register statistics and when micro data are released for research purposes.

Chapter 13   Coordination and coherence

The concept *coherence* refers to that estimates from different surveys can be used together. For example, for a ratio to be meaningful, the numerator and the denominator must be coherent. Consistent surveys give coherent estimates.

Statistics from different sources can be consistent, i.e. have a high level of coherence through:

– ensuring consistency regarding *populations* (relating to definitions of both object and object set),

– ensuring consistency regarding *variables,* and

– using calibration methods that give consistent *estimates*.

An example with business statistics is used to illustrate a method for coordinated and consistent statistics. The aim is to show how inconsistencies can occur between surveys carried out at different points in time.

Chapter 14   Conclusions

In the last chapter we draw some general conclusions from the previous chapters. A new approach towards administrative data is necessary and development of register-based statistics should be recognised as an important field for statistical science.

References

Glossary

Index

CHAPTER 2

How to Structure a Register System

In Chapter 1 the term *register system* is mentioned several times. The two examples in Sections 1.4.1 and 1.4.2, illustrate that the register system plays an important role when new statistical registers are created. In Chart 1.1 we mention that all statistical registers should be included in a coordinated register system. This system will ensure that all data can be integrated and used effectively. In this chapter, the system of statistical registers is described. A statistics producer needs a model to describe the registers that exist in the organisation, and also to describe the links between these registers.

All the registers together can be called a statistical information system, and the understanding of this system is made easier by a conceptual model, which without being too technical describes the system's object types and relations. Models are needed, on the one hand, to describe what the system actually looks like and, on the other hand, to outline a planned improved structure that can be the basis of the development of the system. In this chapter we use the existing register system at Statistics Sweden to explain a register model, whose structure is of general interest even to statistical offices with less developed systems.

First, the general structure of the system is explained. Thereafter, the *base registers* in the system are described, using the registers at Statistics Sweden as illustrations. After a discussion of *standardised variables*, register systems outside Statistics Sweden are mentioned.

2.1   A REGISTER MODEL BASED ON OBJECT TYPES AND RELATIONS

We must first make a distinction between the terms *register* and *register-based survey*. A certain register can be used in many different register-based surveys; it is therefore possible to discuss registers without mentioning a particular survey that will use the registers. In this chapter we discuss registers and the register system in this manner. We here use IT-terms, and statistical terms will be used only when we discuss register-based *surveys*. IT-terms and statistical terms are connected as shown in Chart 2.1.

Chart 2.1 Relations between IT-terms and statistical terms

IT-terms	Statistical terms
Object	Object, statistical unit, element
Object type or object class	Kind of statistical unit
Object set	Population
Relations between objects	Links

Register-based Statistics – Administrative Data for Statistical Purposes   A. Wallgren and B. Wallgren
© 2007 John Wiley & Sons, Ltd

Our work to produce this conceptual model of the register system began with an inventory of all the registers and register products that existed at Statistics Sweden. A statistical survey begins with the definition of the population's objects or statistical units. Correspondingly, the work began with the conceptual model by sorting out the registers in the inventory by type of object. Only statistically important object types were studied, i.e. object types that are included in target populations. Some registers contain hierarchies of object types and it is therefore natural to group them in the following way:

- person and household;
- organisation, enterprise and local unit where an enterprise carries on activities;
- real estate/property, building and dwelling;
- vehicle.

There can be *relations* between different types of objects:

- Between person, enterprise and local unit: a person *is employed* by an enterprise or organisation and *works* at a specific local unit. A similar relation applies between students and places of education: a person *participates in education* that is organised by a specific organisation or enterprise, and the teaching is carried out at a certain place of education/local unit.
- Between person and property/dwelling: a person *is registered* (by the Tax Board) at a specific property or dwelling.
- Between local unit and property: a local unit *is situated* at a specific address and the address relates to a specific property.
- Between person and vehicle, and enterprise and vehicle: a person or an enterprise *owns* a specific vehicle.

These relations are very important from an administrative and legal point-of-view. This is why the administrative sources contain information of good quality about these relations. It is also very important for a functioning register system – when an administrative source contains information on relations, *links* are created in the system.

A link between two objects consists of one or several common *linkage variables* that contain the information needed to identify relations between objects. Because the relations mentioned above refer to relations between different types of objects, these links play a very important role in the register system – in the final model they are the links between the system's base registers. We have created the model below using these objects and relations. The rectangles represent *object sets* and the lines represent *relations* between objects.

Chart 2.2 A conceptual model of Statistics Sweden's register system, version 1

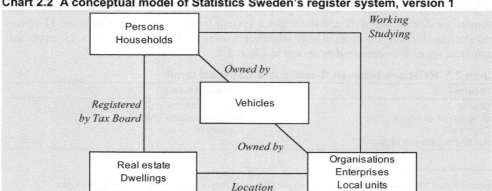

Chart 2.2 shows four registers that contain different types of objects of specific statistical interest. The four rectangles correspond to four Statistics Sweden registers: *the Population, Vehicle, Real Estate and Business Registers*.

When we talk about *the* Population Register, we are actually talking about a series of registers because the Population Register exists in many versions: one with individuals, one with families, different versions for the population at different points in time, etc.

The relation *Registered by Tax Board* is shown in the Population Register by a linkage variable that gives the identity of the residential property where the Tax Board registers a specific person. The relation *Owned by* is shown in the Vehicle Register via the inclusion of the vehicle owner's personal or organisation identity as a linkage variable in the register.

Neither the Population Register nor the Business Register contain linkage variables that show the relation *Working/Studying*. But there are other registers with information on gainful employment or educational activities:

– The Statement of earnings Register (a job register based on advance tax payments) and the Wage Registers contain variables that give information on employment. The object type *employment* is defined by a relation between the object types employed person, employer and local unit. For every combination of personal identity, enterprise/organisation identity and local unit identity, there are data on income/salary, etc.

– The registers with different kinds of students contain variables showing the relation between persons who are studying, course organisers and places of education. Course organisers are organisations/enterprises and places of education are local units.

How should the information on the relation *Working/Studying* be represented in the system? A relation between two object types can be regarded as an object type – a *relational object*. When there are *many* variables describing these relations, it is more convenient to regard them as objects. As there are many statistically interesting variables in different sources describing *Working/Studying*, the best solution is to create a special register consisting of relational objects, i.e. objects that describe the relation between persons, enterprises and local units, in which the same person can have several relations to different enterprises and local units. The reasons behind this are discussed further in Sections 2.2.2, 2.2.5 and 3.3.4.

In Chart 2.3 below, we have introduced the relational object *Working/Studying* as a special register, which we call *the Activity Register*. The relations in Chart 2.2 above have, in the chart below, been replaced by the variables that work as links between objects in the different registers.

Chart 2.3 A conceptual model of Statistics Sweden's register system, version 2

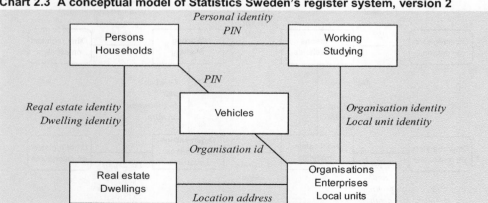

Statistics Sweden's system of statistical registers is based on different administrative sources. The chart below shows the five parts of the tax system that are sources for the five registers shown in Chart 2.3 above.

Chart 2.4 The register system is based on different parts of the tax system

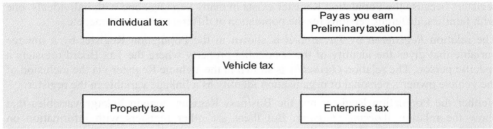

In addition to the five registers shown in Chart 2.3, there are approximately a further 50 registers at Statistics Sweden. Using the existing identifying variables, these can be linked to one of the Population, Activity, Real Estate or Business Registers. These four registers play a particularly significant role within the register system and are therefore called the system's *base registers*. The entire register system is shown in the comprehensive conceptual model in Chart 2.5 below.

The Vehicle Register contains the object type *vehicle*, which has a more limited role for the system as a whole. For the sake of simplicity, we have therefore included individual-owned vehicles in the various registers on individuals and enterprise-owned vehicles in the various registers on enterprises.

The final model is described below. Can this model be used generally? As the same object types and variable content exist in many countries, we believe that this model is generally suitable to describe systems of statistics on society. A national statistical office, with access to the administrative registers mentioned in Chart 2.4 above, can build a system of statistical registers according to the structure described by the final model in Chart 2.5 below.

In the Scandinavian countries the development of register-based statistics started with registers on persons. In other countries it is easier to get access to administrative business registers, and the development starts in that part of the system. However, the final model can be the same in all countries.

Chart 2.5 A conceptual model of a register system of statistics on society

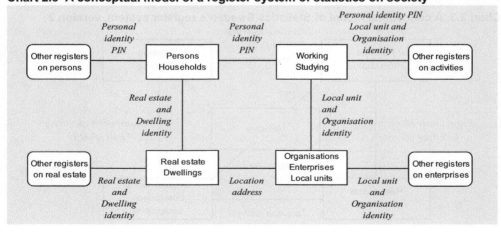

2.2   THE SYSTEM OF BASE REGISTERS

The four base registers, and the links between them, constitute the backbone of the register system, as they contain the important object types and links. If any of these base registers is missing or has a low level of quality, the whole register system would be much less useful for statistical purposes. The role of the base registers is to define objects and populations in the system, where good object definitions and good coverage are important characteristics that are crucial for the quality of the whole system.

The variables that are important in a base register are those that identify an object and that can be used to link that object to objects in other registers. Time information for different events that affect the object is also needed to be able to create populations relating to a specific point in time or period.

In addition to being based on stable and reliable administrative sources, a base register should have the following characteristics:

Chart 2.6 The characteristics of a base register

1. Defines important object types.
2. Defines important object sets or standardised populations.
3. Contains links to objects in other base registers.
4. Contains links to other registers that relate to the same object type.
5. Is important for the system as a whole – which is why it is essential for them to be of high quality and be well-documented.
6. Is important as a sampling frame.
7. Can be used for demographic statistics regarding persons, activities, real estate or enterprises.

    In the same way that age distribution and births and deaths in a population of persons are described, it should be possible to describe age distribution and births and deaths among jobs, buildings or local units. Birth dates and death dates must be available in the base register so that demographic statistics can be produced.

The four base registers are described in Chart 2.7 below.

Chart 2.7 The four base registers

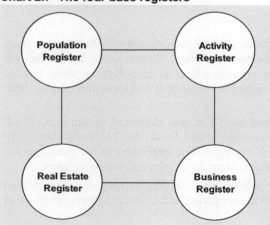

The 'Population Register' in the chart refers to many register versions as each base register can exist in several versions:

– the current population;

– the population at a specific point in time, e.g. December 31;

– all changes during a specific period, e.g. calendar year;

– all objects that have existed during a specific period, e.g. the calendar year register.

We now discuss these four base registers and how the existing registers at Statistics Sweden comply with the seven requirements we have mentioned above.

2.2.1   The Population Register

This base register has advanced the furthest in terms of the development of register-statistical methods for how to work with a base register. It fulfils all of the seven require-ments for a base register outlined in Chart 2.6 above. The Population Register can therefore serve as a source of inspiration for the other base registers. Those working with the Popula-tion Register know how to update, have a long tradition of producing advanced register statistics, regularly perform demographic analyses and are accustomed to creating inte-grated registers for different purposes. An advanced production system has been developed where the histories of the objects are stored.

The register is based on administrative data from the National Tax Board's civic registra-tion system. Statistics Sweden gathers these data daily from the National Tax Board's system, and also receives some annual data from the Swedish Migration Board. The data refer to identification and localisation variables for individuals and demographic variables such as age, sex, civil status, etc.

When the transfer from mainframe computers to a PC network took place at the end of the 1990s, the variable content of the Population Register was increased. Better *address data*, better data on *relations* to other individuals and *dates* for all occurrences in the Population Register present new possibilities for statistical production. Matching is made easier by the registration of amended and temporary personal identification numbers. In addition, differ-ent types of amendments are registered so that it is possible to differentiate between real events and other amendments. These improvements are important for the register's role as a base register. Wilén and Johannesson (2002) have described the new Population Register.

Nevertheless, there are certain negative aspects:

– The definition of households has gaps that could be remedied by the implementation of a good dwellings register and if individuals were registered by dwelling as in Denmark and Finland.
– There is overcoverage for certain categories of immigrants. This was discovered at Statistics Sweden and the National Tax Board was informed. The National Tax Board has since improved its checking of population registration.

2.2.2   The Activity Register

The register consists of three different sections: *gainful employment or job activities, study activities* and *other activities related to the labour market.* As previously mentioned, the register consists of relational objects. Every such object is identified by three linkage variables: personal identification number, organisation number and local unit number from the Business Register.

Data on *job activities* are based on the employer's annual statement of earnings, which gives the income for every combination of *employee* and *local unit.* Tax declarations give information on those who are active as *self-employed.* For employees in the public sector, Statistics Sweden receives monthly information from the employers' salary registers.

The following are characteristics of the *job activities* in the Activity Register: earned income, monthly income, and extent of work, occupation and commuting between place of residence and place of work.

Data on *study activities* are based on data from schools and universities. There are a variety of registers for different kinds of students. These registers contain the students' personal identification numbers and details about the school or place of study, which in turn is a local unit in the Business Register. By giving these places of study a local unit number, it is possible to link the School Register with the Business Register. The activity *studying* in the Activity Register will then be a relation between a personal identification number and a local unit number. Commuting can be identified for students in the same way as commuting for gainfully employed persons, with a personal identification number and a local unit number giving the location of both the place of residence and the local unit.

Data on *other labour market related activities* could also be included in the new register. Different authorities have information on military service, sickness benefits, disability pensions, employment policies, registered unemployment and institutional medical care. This information can give a complementary picture of the status of the labour market over and above information on gainful employment and studies. The administrative sources contain information that can locate activities in terms of time, even if the quality of this is sometimes not so good.

A large amount of administrative data exists containing links to both individuals and enterprises. From a purely technical point-of-view, these data could be considered as data describing relations between individuals and enterprises. As mentioned before, a relation between two object types can be regarded as a relational object. As there are many statistical variables describing these relations, it is more convenient to regard them as objects, which are statistically interesting themselves.

We have chosen to consider this information as data describing *activities*. We see these activities as their own objects for two reasons, firstly because they are a statistically interesting object type and secondly because this object type needs to be distinguishable so that the register system as a whole will have a clear structure. Does the Activity Register fulfil the seven requirements for a base register mentioned in Chart 2.6 above?

1, 2, 7. Defines important object types and object sets, populations. Can be used for demographic statistics regarding activities.
When labour market supply and demand meet, relations are created between individuals and enterprises/organisations. These relations are important for labour market statistics and are described by many important statistical variables. It is convenient to regard these relations as objects. Demographic statistics that describe how, for example, the range of gainful activities changes through *job creation* and *job destruction,* are very relevant in the study of labour market statistics. The Activity Register should therefore contain birth and death times of the activities.

3. Has links to objects in other base registers.
The linkage variable personal identification number is a link to the Population Register and the linkage variables organisation number and local unit number are links to the Business Register.

4. Has links to other registers that refer to the same object type.
The linkage variables personal identification number, organisation number and local unit number are links to the Statement of Earnings Register, the Register of wages[1] and the Occupation Register. Other wage registers and student registers will be standardised regarding the variables that identify local unit and place of study. The Activity Register will then have links to all the registers that contain these relational objects.

---

[1] We use the term *wages* to denote gross wages and salaries

5. Important for the system as a whole.

There is great interest among users for statistics in which data on individuals and data on enterprises are combined. The Activity Register plays a very important role as a bridge between these two types of statistics.

6. Important as a sampling frame.

Certain categories of individuals or enterprises could be selected using the Activity Register. From a register of study activities special categories of students are selected.

The Activity Register can be used to create registers on persons or enterprises

The Activity Register is directly based on administrative sources – for every combination of person and local unit of an enterprise, there are administrative data on annual gross wages. By summing up wage data for every individual, a register on persons can be created. By summing up wage data in the Activity Register for every local unit, we can create a local unit register with data on gross wages ('wage sums' in Chart 2.8).

Chart 2.8 The relation between registers on persons, activities and enterprises

Population Register – Persons		Activity Register – Jobs			
Person	**Wage sum**	**Job**	**Person**	**Local unit**	**Wage sum**
PIN1	450 000	J1	PIN1	LU1	220 000
PIN2	210 000	J2	PIN3	LU1	180 000
PIN3	270 000	J3	PIN1	LU2	230 000
		J4	PIN2	LU2	210 000
		J5	PIN3	LU2	90 000

The Activity Register contains the bi-variate distribution and the Business and Population Registers contain marginal distributions

Business Register – Local units	
Local unit	**Wage sum**
LU1	400 000
LU2	530 000

The chart above describes three statistical registers, which give three different but consistent pictures of society. To understand the register structure, we must be able to distinguish between these three registers and understand how they are related. The Activity Register in Chart 2.8 is neither a register on persons or a Business Register but a register describing relations between persons and enterprises. We will return to this example in Section 3.3.4.

2.2.3   The Business Register

Statistics Sweden receives administrative data concerning *legal units* from the Patent and Registration Office and the National Tax Board. Data from these are received regularly. Extensive work is carried out within Statistics Sweden to collect information from enterprises with activities at more than one local unit, in order to create a register of all *local units* or *establishments*. Within the Business Register, the object types *enterprises, kind of activity units* and *local kind of activity units* are also created.

Chart 2.9 Object types in the Business Register

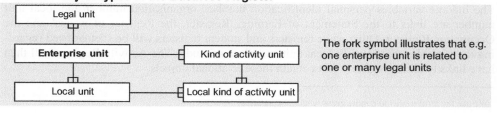

A large number of administrative sources, used by Statistics Sweden, have data about legal units. Statistics Sweden collects data from the other object types. Local units and local kind of activity units are important for regional statistics. Kind of activity units and local kind of activity units are important for economic statistics describing branches of industry. The enterprise units are of central importance as data from all sources, administrative and statistical, can be aggregated and compared for these units.

Two principles are important for internally consistent[2] economic statistics:

- No parallel object sets should appear in economic statistics. So, for instance, the population of energy producers should be the same in the Energy survey, the Business Structure survey and the Business Register.

- Everyone within the Department for Economic Statistics should assist in the maintenance of the Business Register. For instance, those who collect data from manufacturing industries should forward the information they receive on this population to the Business Register.

These principles should apply to all the base registers, not only to the Business Register. Everyone should take part in the maintenance of a base register, regardless of which department they are working in. In practice, this means that those working on the Farm Register, for example, should provide basic information for the agricultural part of the Business Register and that those working with the School Register should provide the required information to make the educational section of the Business Register as good as possible. These principles are discussed in detail in Chapter 5.

Does the Business Register fulfil the seven requirements for a base register mentioned in Chart 2.6 above?

1, 2, 7. Defines important object types and object sets, populations. Can be used for demographic statistics regarding local units and enterprises.

Several important object types and populations for economic statistics are found in the Business Register. Many users request statistics describing the demography of enterprises and local units. Johansson (1997, 2001) includes examples of how the Business Register can be used for demographic enterprise statistics within economic research. The example below also illustrates how important it is to show the changes in the enterprise population. Statistics from the Business Register can end up on the first page of the leading daily newspaper in Sweden:

Dagens Nyheter 12 July 2002:

Here are the new jobs

Would you like a job? Apply to a service enterprise. They are the ones who are employing the most people, according to statistics from Statistics Sweden, produced for Dagens Nyheter ...

Statistics Sweden's statistics should be interpreted with some caution. The enterprises that appear to have grown the most are often subsidiaries of a larger corporate group, where the business has been restructured in some way...

This example shows that statistics based on the Business Register can be newsworthy, i.e. of general interest for a widespread audience. The article refers to the demographics of enterprises. It is also mentioned that the statistics should be interpreted with caution. The

---

[2] When we use the term *consistence* we don't refer to the term used in inference theory. We use the term to explain that different sources give statistical estimates, which are in accordance with each other.

caution is necessary because the administrative data received by Statistics Sweden have in this case not been transformed into statistical data. Processing of the Business Register should be adapted to meet the requirements of the statistics – statistically interesting growth should not be mixed with uninteresting administrative changes such as mergers and acquisitions as in the example above. The Business Register at Statistics Sweden has mainly been used as a sampling frame. Sampling frames are however not suitable for register-based statistics. The demography of enterprises is an important field that has developed strongly over recent years to describe statistically interesting changes in enterprise populations. To meet these needs a different version of the Business Register should be developed. This is discussed in Chapters 5 and 13.

3. Has links to objects in other base registers.
The combination of the linkage variables organisation number and local unit number is a link to the Activity Register. The location address of a local unit is the link to the Real Estate Register.

4. Has links to other registers that refer to the same object type.
Legal units are identified by organisation numbers; the other object types in Chart 2.9 are also identified by number codes, which can be use as linkage variables. The organisation numbers are created by the National Tax Board and are used in all administrative sources. The identification numbers of the other object types are created by Statistics Sweden. For self-employed, the enterprise owner's personal identification number works as the organisation number. All these linkage variables are links to other registers that relate to enterprises/organisations and e.g. local units.

5. Important for the system as a whole.
The Business Register is not only important for enterprise statistics, but also for labour market statistics and statistics on individuals. Many statistical products use data on branch of industry and the location of a local unit.

6. Important as a sampling frame.
Used as a frame for sample surveys concerning enterprises, legal units or local units.

2.2.4   The Real Estate Register

A pure base register for real estate and related object types will be formed. Such a register should be updated often so that divisions and mergers of real estate can be monitored. The register should fulfil our requirements for a base register mentioned in Chart 2.6. The following object types should be included:

– real estate – landed property,
– real estate – buildings and dwellings,
– taxation units and valuation units.

Important variables in such a base register would be the identities of the different objects that belong to the object types listed above. Location addresses, geographic coordinates, and geographic codes should also be included. Addresses should have good quality so that residential real estate and local units of enterprises can be tied to standardised addresses.

Digital maps defining the location of all these objects can be considered as registers and should also be included in the base register. These maps should be created from the coordinates available from the National Land Survey thus defining the geographic dimension in the register system. Today these maps will be maps with coordinate points. If the polygon map of real estates is available this should also be included.

A base register should define populations of objects. Important variables are then those that identify and locate the objects, and define times for important events regarding these objects. Such as register can be built up with the help of the National Land Survey's real estate and building register and be supplemented by a dwellings register and a real estate taxation register. Links to taxation units and buildings should exist in a base register for real estate. The base register could then be updated with data on land registration, data on building permits and real estate taxation data.

2.2.5 Why are there four base registers?

Some countries have a structure with three base registers: individuals, enterprises and real estate. This structure refers to the *administrative* registers that are used to administrate these three types of objects. It is therefore correct to say that there are three important or basic *administrative* registers, which are used for taxation of individuals, enterprises and real estates. But we must differentiate between administrative purposes and statistical purposes – what is a good administrative structure is not necessarily a good statistical structure.

Activities are not administrative objects – the income verifications, which are the basis of the job part of the Activity Register, are used in the taxation of individuals. But activities are interesting statistical objects and there exist many statistically important variables as wage and occupation which describe these activities. As one person can have several jobs and study activities during a calendar year the Activity Register should not be reduced to a register on persons – that would cause a serious information loss. That is why activities are included in a fourth base register in the *statistical* register system.

The conceptual model of the statistical register system must show how the system 'works' – the lines in Chart 2.5 illustrate the links between all registers in the model. These lines or links show how data can be integrated in the system. To be able to show how data is integrated in the system we need a model with four base registers and four basic links – the links between the base registers.

The Activity Register is therefore needed as the link between individuals and enterprises. A register model with four base registers has a clear structure and highlights *activities* as a statistically interesting object set that can be illustrated using a variety of administrative sources. The model is used for the purpose to give a clear picture of the system and to show how the different parts cooperate. A good *statistical* register model should thus consist of four, not three, base registers.

2.3 THE REGISTER SYSTEM AS A WHOLE

The next step is to link the various statistical registers to the respective base register. The circles in Chart 2.10 represent the base registers and the lines show the links between objects in different registers.

A well-functioning system requires that the population and object definitions have been coordinated, that the time references and common variables are harmonised and that there are good identifying variables that can be used to link objects in different registers.

The system in Chart 2.10 below illustrates the content of a well-developed system of statistical registers based on many administrative sources and some censuses.

Some of the registers in the system are created by integration of already existing information in the system. The longitudinal registers are examples of such integrated registers.

Chart 2.10 A system of statistical registers – registers by object type and subject field

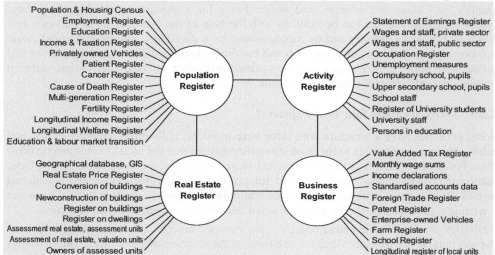

Note: This model is not a model of Statistics Sweden's system, the model shows general possibilities. The Swedish system contains more registers, but Statistics Sweden is not responsible for all of them.

In the real world, objects are born, change and die at all the time. Due to this, the registers in the system are changing constantly – objects are added and taken away. The relations between objects are altered and the properties and variable values of objects change. Every register in the above model exists for several different years and can exist in several versions.

In general, deliveries to the register system can be daily, weekly, monthly or annually. A large part of the Swedish register system receives data monthly or even more often – some register-based statistics can therefore be very timely.

2.4   BUILDING AND USING THE SYSTEM

The part of the register system relating to statistics on individuals is the most highly developed in Sweden and can be used to show the possibilities presented by a coordinated system.

2.4.1   Statistics on individuals based on Sweden's register system

The registers and products shown in Chart 2.11 below illustrate how the system can be created step by step and how it can be used for different applications. The first step is to develop applications using the base register. A base register contains important objects and object sets, and demographic statistics concerning these are an interesting application.

The second step is to develop applications using each administrative source, one application for each main source. The next step is to develop more advanced applications by integrating data from different statistical registers already in the system. Some of these integrated registers can be used for official statistics, and others can be used for academic research.

There are also some special applications using registers. Micro-simulation models are used by researchers and by government and there is great demand for detailed regional tables.

Chart 2.11 Statistics on individuals based on Sweden's register system

1. The base register:	*Demographic statistics:* Population, births/deaths, migration
	Sample surveys: The Population Register is used as sampling frame.
2. Registers directly based on administrative data:	Income & Taxation Register Privately owned Vehicles Patient Register Cancer Register Cause of Death Register
3. Integrated registers for official statistics:	Register-based 'census' Employment Register Education Register
	Sample surveys: Instead of asking questions about e.g. age, income and education, variables from registers in the system are used (registers mentioned under 1, 2 and 3 above). Many variables in the register system can be used as auxiliary variables for stratification and calibration of sampling weights.
4. Integrated registers for research:	Multi-generation Register Fertility Register Longitudinal Income Register Longitudinal Welfare Register Education & labour market transition
5. Micro-simulation models:	The Income & Taxation Register is the basis for a simulation model where planned changes in taxation and transfer payments can be tested. The model is used by government and research.
	The Employment Register is the basis for a regional simulation and forecasting model used by government and research
6. Standardised regional tables:	The Population Register, Income & Taxation Register, Privately owned Vehicles, Employment Register and Education Register are the basis for standardised regional tables used by local government and market analysts.
	The regions can be chosen by GIS technology.

2.4.2   Consistency regarding populations and variables

In Chart 1.1 we give four principles for how administrative data should be used:

1. A statistical office should have access to administrative registers kept by public authorities. This right should be supported by law as the protection of privacy.
2. These administrative registers should be transformed into statistical registers. Many sources should be used and compared during this transformation.
3. All statistical registers should be included in a coordinated register system. This system will ensure that all data can be integrated and used effectively.
4. Consistency regarding populations and variables are necessary for the coherence of estimates from different register-based surveys.

The second principle is illustrated in Sections 1.4.1, 1.4.2 and 1.5.4. The third principle is illustrated in Chapter 2 above. Statistics Denmark has inspired us to develop these two principles. We now explain the fourth principle, where we have been inspired by Statistics Netherlands.

Consistency regarding populations is achieved if all register-based surveys use the same version of the base register as population. The unit responsible for the Population Register creates a special version each year defined as the register describing the population at December 31. This register is used as a standardised population and defines the object set for many other registers with data on persons. This ensures consistency between the different register products regarding the populations.

The responsibility for important variables of common interest is decentralised so that the persons working with the different registers are responsible for 'their' variables. Subject matter expertise is very important during this work, both when the administrative sources are discussed with the authorities that deliver the data and when the administrative variables are processed and statistical variables are created. This is the main reason why it is justified with a decentralised organisation. The chart below illustrates how four different register units at Statistics Sweden are responsible for 'their' standardised variables. Standardised variables are variables that are important for the register system as a whole. This concept is discussed in the next section.

Chart 2.12 Standardised variables – data on persons

Register unit responsible for the ..	Responsibility, standardised variables
... Population Register	Demographic variables, civil status, country of birth
... Employment Register	Status of employment
... Education Register	Final education level
... Income & Taxation Register	Different income variables

The responsibility for variables is decentralised, but the register units cooperate and use the same population. This gives consistent register statistics, which is illustrated by the following table, which contains data from the four registers mentioned above.

Chart 2.13 Register-based statistics for one small municipality in Sweden 2003

Population		Employment		Education					Income Yearly earned, $ thousands				
Age	Number	Em- ployed	Not em- ployed	Com- pulsory	Upper secon- dary	Post- secon- dary	Post- gra- duate	Not known	0	1–14	15–29	30–44	45+
0–15	1416	-	-	-	-	-	-	-	-	-	-	-	-
16–19	387	69	318	306	71	0	0	10	118	265	4	0	0
20–24	293	207	86	44	219	26	0	4	12	130	128	23	0
25–34	764	616	148	79	469	210	0	6	20	133	388	202	21
35–44	937	782	155	142	558	226	2	9	27	128	440	270	72
45–54	1002	847	155	259	510	225	4	4	14	90	501	318	79
55–64	1042	713	329	420	413	199	6	4	21	166	502	288	65
65+	1199	40	1159	333	168	78	3	617	3	552	535	90	19

The four registers in Chart 2.12 above could also be regarded as one large register with all variables, which has been created by a decentralised but coordinated process of work. In Chart 2.13 above the variables are described one by one. But with the large integrated register with all variables, many multi-dimensional tables and analyses could be done. The integrated register would be completely consistent regarding the variables – no conflicting information arises, as common variables in the four registers are identical.

The consistency between statistics from these statistical registers on persons is appreciated by the users, statistics describing about 300 municipalities, 100 age classes and two genders, or 60 000 table cells are completely consistent. An important condition for the possibility to produce coordinated register-based statistics in this way is that the base register, in this case the Population Register, is of high quality.

The integrated register discussed above corresponds to the Social Statistics Database (SSD) discussed by many authors in Statistics Netherlands (2000) and by Laan (2002). The authors of these papers discuss exact matching and micro-integration of data from many statistical surveys: register-based surveys, censuses and sample surveys. They have a vision of complete consistency regarding both population and variables. The Swedish registers we mention above are consistent in these respects and can be integrated at the micro level by exact matching. In Chapter 13 coordination, coherence and consistency are discussed further.

To be able create a coordinated register system where all registers can be integrated, it is necessary that the base registers are coordinated. If calendar-year versions of the base registers are created, then:

- the persons connected with the activities in the Activity Register should also exist in the Population Register;
- the organisations and local units in the Activity Register should also exist in the Business Register;
- the addresses of the local units in the Business Register should be found in the Real Estate Register; and
- the dwellings and real estate where the persons in the Population Register live should be found in the Real Estate Register.

2.5   STANDARDISED VARIABLES IN THE REGISTER SYSTEM

Certain variables are used within many registers and play an important role for the system as a whole. A *standardised variable*[3] is so important that responsibility for the variable is outlined in a special decision. Those responsible for a register in the system, that receives such a variable from an administrative authority or that itself creates such a variable, should have the overall responsibility for that variable at the statistical office. Having overall responsibility for a *standardised variable* involves the following:

- maintaining contact with the authority that provides the administrative register and ensuring that the variable and information on the variable is received by the statistical office from the authority;
- having responsibility for editing, processing, naming and documentation.

Those responsible for other registers at the statistical office that use a standardised variable should use the standardised variable without amendments, its correct name and the original documentation. In this way, duplicate work can be minimised, coordination is made easier and consistency within the system is improved. Two categories of standardised variables have a special role within the system, *links* and *classifications*.

2.5.1   Links

The links that form the basis of the system are shown in Chart 2.5 and are based on the following identities: *person identity number, organisation or legal unit identity number, local unit identity number and real estate and dwelling identity*. In addition to these, *location address of a local unit* is the link between the Business and the Real Estate Registers.

When a register is created in the register system, exact matching with identifying variables is carried out between different registers. An administrative register is matched against a

---

[3] Textbooks use the same term for another concept: A standardised variable has mean 0 and standard deviation 1

base register to check the object sets, and matches are carried out against other statistical registers to gather variables for the new register. All this matching presumes that the different registers contain good links. A link consists of one or several variables that identify separate objects. To facilitate the matching of different registers, it is important that the same identifying variable exists in many registers. When matching a register that relates to different years, links that are stable over time are necessary, i.e. a specific object that is unchanged should have the same values for the identifying variables.

Example: Personal identification numbers were introduced in Sweden 1947. It is a national identity number created by the National Tax Board for every person permanently living in Sweden and registered by the National Tax Board and is used in almost all administrative systems. It is very stable variable with high quality, which facilitates register processing.

Example: The real estate identity in Sweden consists of county, municipality and real estate code. This is an example of a very unstable variable; if the county or municipality code changes due to changed administrative divisions, many real estate identities will also change. It would be much better to use identity numbers, which are kept the same as long as the corresponding real estates remain unchanged.

2.5.2   Classifications

Branch of industry, product category, education, occupation, etc. are examples of important statistical *standards* and *classifications*. These are based on international recommendations, are important in terms of content and are used in many surveys, both register-based studies and others. The administrative sources contain in many cases data on these classifications, and this information is used to create variables within the register system. In addition, those responsible for a standard at the statistical office must be able to *code* new occupation terms, new educational programmes, etc. so that the new terms are included in a suitable category within the classifications.

2.6   STATISTICAL REGISTER SYSTEMS OUTSIDE STATISTICS SWEDEN

Statistical authorities in many countries are becoming more and more interested in basing their statistics on administrative sources. Internal statistics within larger enterprises are also completely register-based.

2.6.1   Register-based statistics in other countries

There is a long tradition of register-based statistics in the Scandinavian countries. When Denmark, Finland and Sweden joined the European Union, the discussion of this 'new' method for producing official statistics started within Eurostat, the organization responsible for the statistical cooperation within the union.

During 1997 a seminar was organised by Eurostat, the proceedings from this seminar (Eurostat, 1997) describes the situation at that time. The seminar was a turning point; afterwards register-based statistics has become a more and more accepted method for producing official statistics within the union. In Section 2.4.2 we mention Statistics Netherlands where they are developing both social and economic databases, to a great extent based on administrative sources.

In their Quality Guidelines, Statistics Canada (2003) distinguishes between different kinds of surveys: censuses, sample surveys and collection of data from administrative records, which we call register-based surveys. The policy of Statistics Canada is to use administrative records whenever this is cost-effective as compared with direct data collection. More and more administrative sources are used, especially for business surveys. In Statistics Canada (2006), the present use of administrative data for enterprises is described. A system approach aiming at consistent economic statistics based on the Business Register is discussed in this paper, which resembles the role we want that base registers should have.

However, administrative data from the Canada Revenue Agency is not used for creating registers, instead this administrative data is used as substitute for sample survey data for small enterprises, and for a fraction of the sample to reduce response burden.

According to the Strategic Plan by the U.S. Census Bureau (2003), respondent burden and cost to taxpayers should be minimized by acquiring and developing high-quality data from sources maintained by other government and commercial entities. One objective is also to produce new information using existing data sources by micro-integration. "One of the greatest opportunities for improving current statistical measures arises from integrating various sources of information." To achieve this goal, research will be conducted to improve methods.

At the U.S. Internal Revenue Service (IRS), tax returns from persons and enterprises are used for statistical purposes. Scheuren and Petska (1993) give an overview of the statistical operations at the IRS. These operations are based on a long and well-established tradition on how to use administrative sources, and we shall here make some comparisons based on our Swedish experiences.

In the model of the register system in Chart 2.10 there are three important registers that correspond to registers maintained by the IRS:

– The Income & Taxation Register on persons and tax family units is based on tax returns from persons. This kind of register is also used for micro-simulation modelling at the IRS and at Statistics Sweden.

– The Statement of Earnings Register in Chart 2.10 corresponds to the 'information documents' mentioned by Scheuren and Petska. Providers of different kinds of income (employers, banks, etc.) report income for each person who gets income. The reports from employers are very important, as this information is the link between persons and enterprises in the register system. As we point out in Section 2.2.2, this information can be used to create the Job Register, which is the major part of the Activity Register.

– Standardised accounts data. This register is based on the tax returns from different kinds of enterprises.

Our conclusion is that the same kinds of important administrative sources exist both in the U.S. and the Scandinavian countries and these sources have been used for statistical purposes for a long time. Due to the size of the U.S., only a sample of all tax returns is used for statistical analysis by the IRS. The sampled tax returns are edited and missing values are replaced with imputed values. In Sweden, sampling of administrative records is used for the micro-simulation model, which is described in Section 6.3.1. We believe that when computers become better and better, the practice of sampling will be considered unnecessary and statistical income registers based on tax returns will be common practice.

Scheuren (1999) discusses the use of administrative records in census applications. He compares the conditions in the U.S., Canada and Europe, where the challenge is greatest in the U.S. It is clear that many administrative sources exist in the U.S. so that a census can be partially register-based. However, there must be laws that require that agencies cooperate, like the Statistics Act in Canada. The public must also get accustomed to the idea that interviews and questionnaires are replaced by linking of administrative records.

In a series of work sessions organised by United Nations Economic Commission for Europe (UN/ECE) and Eurostat, register-based population and housing censuses have been discussed. In UN/ECE (1998) recommendations for the 2000 censuses are given. The method of using registers and other administrative sources, if necessary in combination with sample surveys, is a method that is accepted by UN/ECE.

The Population and Housing Census of 1981 in Denmark was the first complete census in the world based exclusively on information from administrative registers. Register-based surveys have for a long time been considered as the most important and cost effective statistical methodology by Statistics Denmark. The first book on register-based surveys was published during 1994 (in Danish) followed by a version in English, Statistics Denmark (1995).

Statistics Finland (2004) gives a description of the register-based statistics in Finland. Since 1990 Statistics Finland has been able to produce completely register-based population and housing censuses.

In the chart below we indicate which parts of the register system in Chart 2.10 that are used in a register-based population and housing census.

Chart 2.14 Statistical registers used in a register-based census

Census registers	Register used to produce the census registers
Population Register	
Employment Register	Statement of Earnings Register (Job register based on advance tax payments) Business Register
Education Register	Compulsory school Upper secondary school Register of University students
Income & Taxation Register	
Register on buildings	
Register on dwellings	

As we have mentioned earlier, Statistics Netherlands has developed a Social Statistics Database or SSD. Many administrative sources are used to create statistical registers, which correspond to the registers mentioned above. The Dutch virtual census 2001 was based on the Social Statistics Database, which in turn is based on a combination of registers and sample surveys.

The sources used, and an outline of the methodology used in the virtual census is described in Statistics Netherlands (2004). In Chapter 4, we describe this combination of registers and sample surveys.

Bethlehem et al. (2006) gives a description of the Dutch strategy to change from sample survey-based statistics to register-based statistics. They note that customer demands are changing so that there is a growing demand for more thematic publications in which data from various sources are combined. There is also a growing demand for detailed regional statistics. Register-based surveys where data from many sources are integrated can meet these demands. Bethlehem et al. also describe a model of the Dutch register system.

Important administrative sources

Statistical offices in other countries often work with surveys that concern the same object types as Statistics Sweden's register system. This means that it would be possible to structure their existing or planned register systems in the same way as Statistics Sweden's system, as shown in Charts 2.5 or 2.10. The main administrative sources behind such a system are shown in Chart 2.4:

– income tax of persons including that persons are registered at an address, real estate or dwelling;

– property tax of land and buildings;

– income tax and goods and service tax (or value added tax) of enterprises; and

– administrative sources linking employers and employees.

It is often the case in many countries that the statistics produced are traditionally structured in two sub-systems, one for individual or social statistics and another for economic statistics. Therefore, the possibilities for integrating information from registers on individuals and registers on enterprises are often overlooked.

With a register system as described in Chart 2.10, it is possible to describe *persons* through the characteristics of the *enterprises* in which they are employed – employed persons by industry, by enterprise size, by enterprise age, etc.

Correspondingly, the *enterprise* can be described through the characteristics of the *persons* who make up the enterprise's staff – proportion of women/men, proportion of young persons, proportion of highly educated, etc.

To meet the need for statistics of this kind, registers on persons and registers on enterprises should be incorporated in *one* coordinated system.

National identity numbers and other identities

Statistical offices in the Scandinavian countries have the advantage that many nationwide administrative systems since many years use unique national identity numbers. All systems use the same unique PIN for persons and the same unique BIN for enterprises (legal units). There are also unique address code numbers and real estate identity numbers. All these national identity numbers facilitates an efficient use of administrative data for statistical purposes.

However, these identity numbers are not as perfect as statisticians outside Scandinavia may think. Two objects can have the same identity number; the same number can be used by two objects, etc. This means that matching errors and problems always exists to some extent. As a rule, the matching problems are troublesome when a new register has been created or when new administrative sources are used for the first time. These problems will become less disturbing later – the work with quality improvements is a long-range project that gradually will reduce the matching problems.

Example: In the first version of the Swedish Employment Register 1985 it was possible to link 93.6% of the persons employed with the local unit in the Business Register. After nine years it was possible to link as much as 98.3% of the persons employed.

In countries without national identity numbers, other kinds of identities should be used in the statistical register system. Name, address, birth date, birthplace, etc. are possible linkage variables that can be used as links. The national statistical office should keep track of changed names and addresses to be able to use such variables as linkage variables. The statistical office could also create their own identity numbers based on available linkage variables. These numbers are then used as links in the final versions of all statistical registers belonging to the register system.

Matching within the system is done with the links in the system. The purpose of the matching process is to find identical objects in different registers or to find objects that have a defined relation. These links have been created with available identifying variables.

Different countries can have different identifying variables when these links are created and the matching problems will differ to some extent, but the same theory for statistical register systems and register-based surveys can be used.

2.6.2   Register-based statistics within enterprises

Enterprises and organisations have their own administrative register systems that can also be used for statistical purposes. Enterprises in the manufacturing industry have administrative systems consisting of two main parts: a financial system and a system for material and production management. These systems contain hundreds of registers with thousands of variables.

In the same way as a national statistical office uses administrative registers to create statistical registers, these enterprises can use their administrative registers to create registers that are then used as sources for the enterprise's internal register-based statistics e.g. on sales. These statistical registers contain microdata for all the transactions relating to new orders and invoicing.

In the register system of a manufacturing enterprise, certain registers can be considered as base registers. The items register and the client register are two examples that define important object types. Important linkage variables in the system would then be client identity and item identity.

As we describe in this book how society's administrative systems can be used for statistical purposes, it could be possible to investigate how an enterprise's administrative data can be used statistically. Statistical science should contribute to the development in this field. The need for this is illustrated by the quick growth of 'Data Mining'.

We will illustrate this by an example describing a register-based survey in a manufacturing enterprise. The survey in the example is the monthly survey on sales. Every month a statistical register is created by matchings and selections from three administrative registers, the Invoice Register with all transactions regarding invoices, the Client Register and the Item Register.

Chart 2.15 Three administrative registers

Invoice Register

Date	Client number	Item number	Quantity	Value
2006-01-18	196	22	10	832
2006-01-19	28	4	500	20339
2006-01-19	7	128	40	9840
2006-01-20	23	9	100	10622

Client Register

Client number	Seg-ment	Coun-try
7	3	SE
23	3	SE
28	3	SE
196	2	GB

Item Register

Item number	Item group	Pre-calcu-lated cost
4	1	36
9	1	90
22	2	28
128	2	205

A statistical Sales Register is created in the following way:

From the Invoice register all transactions for one defined month are selected to be the objects in the Sales Register. This monthly register is matched against the Client and Item registers and variables from these registers are imported into the Sales Register.

Chart 2.16 Sales Register for January 2006 – four transactions

Date	Client number	Seg-ment	Coun-try	Item number	Item group	Quantity	Value	Price	Pre-calcu-lated cost	Gross profit
2006-01-18	196	2	GB	22	2	10	832	8.32	280	552
2006-01-19	28	3	SE	4	1	500	20339	40.68	18000	2339
2006-01-19	7	3	SE	128	2	40	9840	246.00	8200	1640
2006-01-20	23	3	SE	9	1	100	10622	106.22	9000	1622

The Sales Register for a given month is then used to create tables with invoiced values at current prices, price indexes, invoiced volumes (values at constant prices) and gross profit margins by segments, countries and item groups. These tables are used to update a time-series database.

Administrative registers must be transformed into statistical registers. In this case the administrative data can be of good administrative quality, but still be unsuitable for statistical purposes.

Two examples below illustrate the importance of transforming administrative data – errors in administrative data should be corrected according to statistical principles, and missing values should be treated as missing values in statistical data.

1. Corrections in accounting data

At February 27, an invoice transaction is registered in the Invoice Register. When the transactions for February are checked it is realised that this transaction is wrong and should not have been done. A correction is done at Mars 2.

Chart 2.17

Administrative Invoice Register

Date	Client number	Item number	Quantity	Value
2006-02-27	53	9	1000	107560
2006-02-28	34	112	655	32700
2006-03-01	117	4	500	20339
2006-03-02	53	9	–1000	–107560

Statistical Invoice Register

Date	Client number	Item number	Quantity	Value
2006-02-28	34	112	655	32700
2006-03-01	117	4	500	20339

According to accounting principles, errors are corrected by adding a new transaction so that the erroneous transaction and the correction will sum up to zero. From a statistical point of view this corresponds to two errors with different signs. In the statistical register neither the error nor correction should be included as the time series then will be of low statistical quality when the error and the correction appears during different months.

2. Missing values

Administrative registers can contain missing values. If these are interpreted as zeros, the statistical analysis will be misleading. When the statistical register is created, missing values should be detected and replaced by imputed values.

Chart 2.18 Administrative Item Register

Item number	Item group	Pre-calcu-lated cost	
4	1	36	It may be that pre-calculated costs have not been calculated for earlier months.
9	1		If these missing values are not detected the gross margin trend will be wrong.
22	2	28	Administrative data must be edited before it is used for statistical purposes.
128	2	205	

Chart 2.19 Statistical Item Register

Item number	Item group	Pre-calcu-lated cost	Total sales of items with known pre-calculated costs during the period is 90 000 SEK
4	1	36	Pre-calculated costs of these sales is 80 000 SEK
9	1	119.5	The average price of item 9 during the period is 106.22 SEK
22	2	28	The imputed pre-calculated cost for item 9 is then:
128	2	205	$106.22 \cdot (90\ 000/80\ 000) = 119.50$

Register-based surveys are common within enterprises and other organisations. In the same way as we in the rest of this book discuss how statistical offices to produce official statistics can use administrative data, the methodological problems associated with these enterprise surveys should be investigated and discussed. It is often the case that these problems are not recognised as statistical problems due to the fact that statistical science is only associated with survey sampling, randomised experiments, probability and inference theory.

CHAPTER 3

A Terminology for Register-based Surveys

The development of register-based statistics requires a common and rich register-statistical language. A common language within the theory of survey sampling is taken for granted. Terms such as sampling frames, probability sampling, estimators and standard errors are well known and have a clearly defined meaning. Register-based statistics have the same need for well-established terms. We think with words and, if our register-statistical terms are unclear, the thinking also becomes unclear. If we, instead, have a range of good and well-defined concepts, the exchange of knowledge and methodology development will be stimulated.

All register-statistical terms with explanations are compiled in the Glossary at the end of the book.

3.1 TERMINOLOGY – DIFFERENT LANGUAGE

Register processing has been influenced by IT theory and IT terms have frequently been used while statistical terms have been neglected. Sometimes terms even have different meanings; when they are used in an IT context, they mean something different from when they are used in a statistical context. The aim with the register-statistical terminology introduced in this book is to name important concepts and introduce unambiguous terms so that a common register theory can be developed.

3.1.1 Concepts and terms

A *concept* is the abstract content of a linguistic *term*. Different terms can exist that refer to the same concept. Some examples are shown below of how statistical terms and IT terms can be linguistically different despite referring to the same concept.

The rapid development in the field of IT has meant that certain specialist terms are changed when the technology changes. This can cause confusion and it may not be realised that the same statistical concept is being referred to. When Statistics Sweden migrated from mainframe computers to database servers, old terms such as *flat file* with *records* and *positions* were replaced by the term *database table* with *rows* and *columns*. Register-statistical terms that survive such technical changes are necessary. That is why we use the term *data matrix* instead of terms such as *worksheet, data set, database table* or *flat file*.

Register-based Statistics – Administrative Data for Statistical Purposes   A. Wallgren and B. Wallgren
© 2007 John Wiley & Sons, Ltd

3.1.2   The term 'statistical survey'

Section 1.2 discusses different types of statistical surveys: sample surveys, censuses and register-based surveys, which form the basis for the statistical production at a statistical office. But the term 'survey' is often used to describe a sample survey. An example of this is the ASA-definition[1] where the term survey clearly means sample survey. There are also a number of (good) books where terms like 'Survey Errors', 'Survey Methodology' and 'Survey Quality' only refer to sample surveys.

Why should register-based surveys also be called statistical surveys? Because the theory should also here be based on general statistical survey methodology. With register-based surveys, it is also necessary to define the research objectives, define population and variables, procure data (from another authority or from the register system of the statistical office), and work with quality assurance, analysis and presentation. By using the term *survey*, these similarities are emphasised.

3.2   REGISTER TERMS

Register and *table* are two important terms that can have different meanings, although it is advisable to ensure that there can be no misunderstandings when using these terms. The term *register* was introduced in Section 1.3, where it had the following definition: A register aims to be a *complete list* of the objects in a specific object set or population and it should contain information on the object's *identity* so that it can be updated with new variable values for that object. This definition applies to both administrative and statistical registers.

3.2.1   Register, data matrix and table as statistical terms

We discuss the terms *register, data matrix* and *table* using the following example with data from an imaginary statistical register. Imagine that we have a register containing data on all enterprises in the country at a certain point in time. The number of objects in the register, illustrated in the chart below, is given by N and the register contains six variables.

Chart 3.1 Example of a register and data matrix

	Variable 1 Name of enterprise	Variable 2 Address	Variable 3 Organisation number[2]	Variable 4 Turnover	Variable 5 Employees	Variable 6 Industry
Object 1	A's Painters	Address 1	BIN 1	12	9	F
Object 2	B's Bakery	Address 2	BIN 2	3	4	D
Object 3	C's Salon	Address 3	BIN 3	7	10	O
.	.	.	.	.	.	.
Object N	Z's Factory	Address N	BIN N	211	76	D

In this simple example, the data matrix and the register correspond. This is usually not the case. These concepts are also compared in Section 3.2.2 below.

---

[1] American Statistical Association (1996): 'A survey gathers information from a sample'
[2] BIN = Business Identification Number

In a *data matrix*, statistical data are sorted so that the matrix columns are the variables and the matrix rows are the observations for the objects. The register in Chart 3.1 is represented by a data matrix with N rows and six columns. The data matrix has been marked out in a box, and we have also added explanatory text in the first column and the heading.

Every statistical survey aims to create one or several data matrices containing *microdata*, which will then be processed for statistical purposes. The term *data matrix* can be considered a statistical concept for such a data set.

The columns in the matrix contain *measurements*[3] of *variables*[4]; the rows in the matrix contain *observations*[5] for the objects in the register. The observation for Object 2 has been marked in white in the chart. The observations are, in this case, six-dimensional. If we receive new data with, for example, an organisation number and a revised turnover, the register in Chart 3.1 can be updated by forming a new version in which the old values are replaced by the new.

A data matrix contains observations for *individual* objects, which is called microdata. If the register is anonymised, we get the data matrix in Chart 3.2 below. This data matrix does not contain any data on the identity of the objects and it is not possible to add new measurements. It is therefore *no longer* a register. The expression *anonymised register* is sometimes used. This is not really appropriate as it is contrary to the definition of the term register.

Chart 3.2 Anonymised data matrix

	Variable 1 Turnover	Variable 2 Employees	Variable 3 Industry
Object 1	12	9	F
Object 2	3	4	D
Object N	211	76	D

A sample from the register in Chart 3.1 above gives a third data matrix and; as the sample is not a complete list of the population, this data matrix is not a register in our sense of the word.

In data matrices for samples, every observation in the sample can be considered as representing many observations in the population. In Chart 3.3, the first observation represents 411.7 observations in the population. An important part of the work with sample surveys is to calculate these *weights*.

Chart 3.3 A data matrix from a sample survey

	Variable 1 Name of enterprise	Variable 2 Address	Variable 3 Organisation number	Variable 4 Turnover	Variable 5 Employees	Variable 6 Industry	Variable 7 Weight
Object 1	X Ltd	Address 1	BIN 1	111	57	D	411.7
Object 2	Y company	Address 2	BIN 2	1	2	G	823.2
Object n	Z Ltd	Address n	BIN n	56	38	F	411.7

By processing the data matrices shown in Charts 3.1, 3.2 or 3.3, it is possible to create tables such as the one shown below. By grouping the variable values, *grouped variables* and *variables divided into class interval* are created.

---

[3] Measurement: measured value for a variable of a specific object
[4] Variable: measurable characteristic of an object
[5] Observation: all measurements for a specific object, also called record

In the two-way table shown below, there are two *spanning variables*: industry and number of employees. The variable *industry* has been grouped and the variable *number of employees* has been divided into class intervals.

Chart 3.4 Example of a table created with data from Charts 3.1, 3.2 or 3.3
Enterprises by industry and number of employees 2000
Number of enterprises

Industry	Number of employees					
	0	1–9	10–49	50–99	100–199	200–
A–F, primary goods	198 006	45 124	8 936	1 144	587	626
G–K, private services	231 910	107 296	14 830	1 424	645	568
L–Q, public sector services	65 783	24 621	3 713	434	200	548
Unknown	105 823	2 515	0	0	0	0

The table's cells are defined by combining the spanning variables. Observations are distributed so that every observation belongs to only one cell in the table. The value in a table cell gives a summary description of the cell's observations. Such summary descriptions of groups of observations are called *aggregated data* or *macrodata*. By using the same data matrix, it is possible to form several tables with different content.

A data matrix can be stored in a database table, which is often called table. Unfortunately, this means that there is one term for two different concepts. We will use the term *table* to denote the statistical concept only.

3.2.2 The terms database and database table

Data matrices such as those shown in Charts 3.1–3.3 are stored in databases, which use various IT terms that we will describe in this section. Section 2.1 describes conceptual models but these often differ from the physical implementation of a database. In this book, we are always talking on a conceptual level, i.e. how it is *thought* that registers and data matrices will 'look'. Chart 3.1, for example, shows complete agreement between the register, the data matrix and the database table but, in an actual IT solution, this may not be the case.

The aim with a systems solution is that the database should be a flexible base for many uses in the register system. The structure of the data matrix can also be saved in a so-called 'view'. IT-terms as databases, database tables and views are discussed in Chapter 11.

Chart 3.5 illustrates the different ways of describing the same database. The part A of the chart shows the traditional way of describing a database. The part B shows an example of a database's contents. The chart shows the structure of the systems solution, i.e. how the data is physically stored. The database is normalised to ensure optimum consistency and space-effective storage.

Chart 3.5 A database on individuals with three database tables
A. Conceptual database model

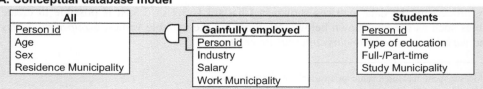

Chart 3.5 A database on individuals with three database tables
B. Example of content in the database

All

Person id	Age	Sex	ResMun.
PIN1	20	F	0586
PIN2	23	M	0586
PIN3	31	M	0586
PIN4	32	F	0586
PIN5	33	M	0586
PIN6	40	F	0586
PIN7	59	F	0586
PIN8	65	M	0586
PIN9	71	F	0586

Gainfully employed

Person id	Industry	Salary	WorkMun.
PIN2	G	52 000	0586
PIN3	G	287 000	0580
PIN4	A	193 000	0586
PIN6	D	291 000	0586
PIN7	D	314 000	0580

Students

Person id	Educ.Type	Full/Part-time	StudMun.
PIN1	AdultEduc	100	0586
PIN2	Univ	100	0580
PIN5	Univ	100	0580

With the above database, it is possible to carry out statistical processing for different aims. Chart 3.6 below shows an example of two different data matrices that could be created from this database. These data matrices can be created and stored physically but they can also be formed temporarily during processing. A conceptual data matrix focuses on the specific problem to be solved, i.e. the statistical analysis that is to be carried out.

Chart 3.6 shows an example of two data matrices for employment and commuting. The data matrices in the chart would, in database terminology, is called denormalised, as several of the cells are missing values.

Chart 3.6 Two data matrices for different statistical purposes

A. Data matrix: Employment register

Person	Age	Sex	Emp-loyed	Industry	Salary
PIN1	20	F	No	null	0
PIN2	23	M	Yes	G	52 000
PIN3	31	M	Yes	G	287 000
PIN4	32	F	Yes	A	193 000
PIN5	33	M	No	null	0
PIN6	40	F	Yes	D	291 000
PIN7	59	F	Yes	D	314 000
PIN8	65	M	No	null	0
PIN9	71	F	No	null	0

B. Data matrix: Commuting register

Person	ResMun.	WorkMun.	StudMun.	Com-muting
PIN1	0586	null	0586	0
PIN2	0586	0586	0580	1
PIN3	0586	0580	null	1
PIN4	0586	0586	null	0
PIN5	0586	null	0580	1
PIN6	0586	0586	null	0
PIN7	0586	0580	null	1
PIN8	0586	null	null	0
PIN9	0586	null	null	0

The IT term *table* is many times used instead of the complete term *database table* (actually relational table). Because the statistical term *table* refers to a method of presenting aggregated data, *macrodata,* this choice of language can lead to misunderstandings when talking to both statisticians and IT experts. Database tables should not be mixed up with statistical tables.

What do we mean by the term *register?* The example in Charts 3.5 and 3.6 refers to whole populations, not samples. The object's identity is included, which means that both Charts 3.5 and Chart 3.6 show registers in our sense of the word. We can also call the database in Chart 3.5 a register, containing several sub-registers. When the register is documented, the systems solution in Chart 3.5 should be given, plus the statistical and content-related data matrices in Chart 3.6. Below, the terms *register* and *data matrix* are summarised.

Chart 3.7 What do we mean with the terms Register and Data matrix?

Register	Data matrix
Consists of all objects that belong to a specific defined object set or population	A collection of microdata to be used for specific statistical processing
Often consists of many subregisters and register versions	Contains primarily statistical variables
The objects can be identified	Can be anonymised
Contains several types of variables, both statistical variables and other	Can come from a sample survey, census or register-based survey
Stored in one or several databases	Can be stored in a database table but can also be created at time of calculation

Terminology also needs to be understood by the public

Internally, within a statistical office, it can happen that the term *register* is used with different meanings. The term *table* can also relate sometimes to a database table and sometimes to a statistical table. The terms that are used externally, on websites and in publications, need to be understandable for the general public.

The general public probably interpret the term *register* to consist of the entire population – a membership register, for example, contains all members, not just a sample. The term *table* is also probably linked to a statistical table by those using the website of a statistical office. In the external communications, the terms register and table should therefore always be used in a way that is consistent and, for the users, easy to understand.

3.2.3   Terms for different kinds of registers

The statistical office receives *administrative registers* and processes these to create *statistical registers*. The four statistical registers that are most important for the register system are called *base registers*, as described in Section 2.2. This term should only be used for these four registers. The base registers are based on administrative sources and with these sources the *object sets* and *object types* of the register system are created.

The other statistical registers (except the base registers), are divided into two categories, *primary registers* and *integrated registers*. We refer to statistical registers that are *directly* based on at least one administrative source as *primary registers*. The primary registers are based on administrative sources and with these sources the main part of the *statistical variables* of the register system are created.

Integrated registers are statistical registers that have been created by *only* combining information that already existed in the statistical registers in the system. This term aims to emphasise this important usage of the register system of carrying out new surveys using existing data only.

Information can be combined from several registers to create a new register without this being an integrated register. For instance, the Income and Taxation (I&T) Register is created by integrating six administrative sources and five statistical registers (Chart 1.4). However, the I&T register is a primary register because it is directly based on at least one administrative source.

A *longitudinal register* is a special type of integrated register where integrated information from several annual registers is compiled so that it is possible to follow identical objects over time.

Chart 3.8 Different kinds of registers

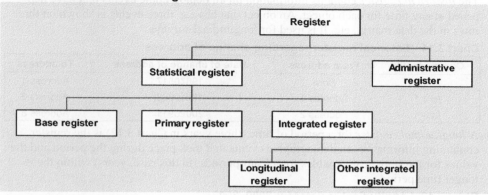

3.2.4 Registers and time

Individual objects, and therefore also object sets, change over time. Objects are born, change location, are altered or cease to exist. These different types of occurrences are called *demographic events* and it is these that change object sets. When defining a register with regard to *time*, the following register types must be distinguished:

1. The *current stock register* is the register version that is updated with all available information on currently active/live objects. The current stock register is used as frame population for sample surveys or censuses.

2. The *register referring to a specific point in time*, such as the turn of the year, is the version of the register that is updated to describe the object set at that point in time. This update is carried out *after* the point in time, when information on all events up to that point in time is available. Is used for register-based surveys.

3. The *calendar year register* is the register version containing all objects that have existed at any point during a specific year. Objects that are added or cease to exist during the year are included with information on the date of the event. It is used as register populations for register-based surveys.

Chart 3.9 Calendar year register for 2002

Object identity	Existed 1/1	Added	Ceased to exist	Existed 31/12	Other variables
Idnr 1	Yes	-	20020517	No	...
Idnr 2	Yes	-	-	Yes	...
Idnr 3	No	20020315	20020925	No	...
Idnr 4	No	20020606	-	Yes	...

4. The *events register* for a specific period, is the register containing information on all demographic occurrences that have taken place during the period. A register is created for every type of event. It is used in register-based surveys.

Chart 3.10 Events register for 2002 regarding change of address

Object identity	Address 1/1	Date of change of address	New address
Idnr 1	Address 11	20020517	Address 21
Idnr 2	Address 12	20020606	Address 22
Idnr 3	Address 13	20020911	Address 23

5. The *historical register* contains information on all demographic events that have happened at any time for each object. An object that has e.g. three events is shown on three rows in the data matrix, etc. It is used for longitudinal surveys.

Chart 3.11 Historical register regarding change of address

Object identity	From address	Date of change of address	To address
Idnr 1	Born	19670517	Address 1
Idnr 1	Address 1	19810606	Address 2
Idnr 1	Address 2	20020911	Address 3

6. A *longitudinal register* for a period of time (three years in Chart 3.12) is the register containing information on demographic events that took place during the period and the values for the statistical variables for all sub-periods (in this case, years) within the longer time period.

Chart 3.12 Longitudinal register for 2000–2002

Object identity	Existed 1/1/2000	Added	Ceased to exist	Income 2000	Income 2001	Income 2002
Idnr 1	Yes	-	20010517	183 450	97 600	-
Idnr 2	Yes	-	-	273 500	281 360	258 340
Idnr 3	No	20010315	20020925	-	193 570	204 520
Idnr 4	No	20020911	-	-	-	56 300

3.2.5   The term 'object'

We have, for the most part, used the term object. There are a variety of concepts and terms that go together with this term.

Population, object, object type, object instance, class, unit, element

'The population U consists of N elements/units/objects.' Discussions in survey theory often begin in this way. Textbooks on statistics most often use the terms elements or units but the term object is common at statistical offices. The corresponding term in the conceptual modelling of an IT system is *object instance.* This terminology refers to a population of individuals of *object type* person, where every individual in the population is an object instance. The term *object class* is often used, primarily in object-oriented programming, as a synonym of object type.

To define a population, the object types must be defined, i.e. what is meant by household, local unit, etc. The set of objects to be included in the population must also be defined and, in this definition, place and time information should always be included, such as households in a certain municipality at a certain point in time.

The term 'object' in statistics and IT

Statistical science uses the term unit or object only for those objects that a statistical survey relates to. Section 2.1 gives the various object types that occur in a statistical office's register-based surveys:

– person and household;
– organisation, enterprise and local unit;
– real estate/property and dwelling;
– vehicle;
– activity, gainful employment and studies.

In an IT environment, the term object is used frequently and sometimes with definitions that differ from the statistical term. In a database solution, rows in certain database tables are called objects without being objects in the statistical or conceptual meaning. This can cause misunderstandings. When a survey is documented, only objects that are part of the register population should be called objects in the statistical part of the documentation. The part of documentation describing the IT system can contain other systems-related objects.

3.3   TERMS FOR DIFFERENT KINDS OF VARIABLES

Columns in a data matrix contain measurements for different variables. Variables can have different attributes and be used in different ways in the statistical processing. Variables that are part of a register system can have been created in different ways and have special functions. Variables are discussed both within statistical science and informatics.

3.3.1   Variables in statistical science

A variable is a measurable attribute of an object. When collecting data, we receive values that are measurements of these attributes. It is these measurements that are compiled in a data matrix. We must theoretically differentiate between the characteristics of an object and the measurement we have collected.

> A statistical variable is defined by which *object type* has the characteristic (e.g. income for persons and income for households are two different variables), by the *measurement method* and *scale* used and also by the given *point in time or period* that the measurement refers to.

Example: The characteristic 'age' of an individual at a specific point in time can be measured by judgement (young, middle-aged or old) or by asking about the date of birth and then calculating the age.

A variable can be *quantitative* or *qualitative*. Characteristics that can be described as a number (such as the age of a person) can be measured with both quantitative (such as age in years) and qualitative variables (such as young, middle-aged or old).

Characteristics that show a category (such as sex) can only be measured with *qualitative* variables, even if codes are used to denote the different categories (such as 1 for male and 2 for female). A qualitative variable is formed by the classification of objects, i.e. when they are divided into a number of groups, which is why such groups are sometimes called *classifications*.

Certain qualitative variables contain many categories, which are also sorted in a *hierarchical* way, such as industry for enterprises. At the highest level in a hierarchical classification (one digit level), the breakdown is crude, becoming finer at the two-digit level and so on.

For *quantitative* variables, the values can be used in calculations and, in a data matrix; these values in a variable column can be aggregated into sums and means. With qualitative variables, it is possible to calculate the *number* with a certain code, such as the number of 1s or 2s in the data matrix column with the variable sex.

> A *qualitative* variable can be transformed into *quantitative* variables using 0-1 coding. E.g. sex can be transformed into two 0-1 coded variables:
>
> $x_1 = 1$ for female,   0 otherwise $x_2 = 1$ for male,   0 otherwise
>
> x_1 is used to describe the number of women and x_2 is used to describe the number of men
>
> For both x_1 and x_2, the mean value and standard deviation can be calculated

Variables, registers and time

In Section 3.2.4 six different register types are discussed where time is treated in different ways. There are two types of variables where time also is treated differently. The combination of variable type and register type is important and should not be overlooked.

- *Flow variables* show sums for different time periods, e.g. earned income during a year for a person, new orders during a month for an enterprise.

- *Stock variables* give the situation at a specific point in time, e.g. age of an individual at a specific point in time or number of employees in an enterprise at the end of the year.

A flow variable should be defined for a calendar year register (or a register version created for a quarter or month). A register with earned income of persons during a certain year should consist of all persons belonging to the population at least some part of that year. The present version of the Swedish Income & Taxation Register (I&T) describes the income during year t of persons permanently living in Sweden on December 31, year t. However, there are persons who were permanently living in Sweden earlier than December 31, but who had left Sweden or died before December 31. In a calendar year register for year t, there incomes would have been included. In the present version of the I&T total income is smaller than the total income in a calendar year register. In a corresponding way, a stock variable should be defined for a register referring to a specific point in time.

3.3.2   Variables within informatics

A variety of terms are used in the field of IT to distinguish different types of variables. Unfortunately, these terms are not usually mentioned in statistics literature but, because the terms are important, particularly in a register-statistical context, we mention a few of them here.

The set of values that a variable can take on, or can be thought to take on, for any object is called the variable's *value set. Example:* The quantitative variable *age of an individual* has the value set 0–115 years.

Example: The qualitative variable *county of residence for an individual* (in Sweden) has the value set Stockholm County, …, Norrbotten County and the qualitative variable *county code* has the value set 01, 03, …, 10, 12, 13, 14, 17, …, 25.

A *single-valued variable* is a variable that takes on only *one* value for each object. In general, variables are single-valued. *Example:* Age of an individual.

Multi-valued variables can at the same moment take on *several* values for certain objects. The number of values differs among objects. *Example:* Industry of an enterprise; one enterprise can carry out activities in several sectors at the same time. The number of industries can vary between enterprises. Multi-valued variables give rise to many important methodological issues that are discussed in Chapter 9.

3.3.3   Derived variables

Derived variables play a central role in register-based surveys. When we collect data in sample surveys or censuses, we design the questionnaire with the questions that define the variables of the survey. In register-based surveys we don't have this opportunity, instead of designing questionnaires we must create derived variables using all available administrative variables. Derived variables are also discussed in Section 6.2.

When we process statistical data, an important part of the work consists of creating derived variables. These can be formed using variables defined for the objects in the register in question. Derived variables can also be formed by using variables defined for other objects. In this case, matching different registers will be necessary. Four different types of derived variables can be differentiated as described below.

1. Variables derived by grouping values and dividing into class intervals

PIN	Age, years	Age class	Country of birth	Geographical category
1	76	70–79	Sweden	Sweden
2	49	40–49	France	Europe
3	32	30–39	Norway	Nordic
4	11	10–19	Chile	Outside Europe

A *quantitative* variable, as age, can be divided into *class intervals*, e.g. 0–9, 10–11, … 80–89, 90 and older.
A *qualitative* variable, as country of birth, can be *grouped* into broader categories, e.g. Sweden, Scandinavian countries, Europe and Outside Europe.

2. Variables derived by arithmetic operations using variables in the data matrix

It is possible with quantitative variables to carry out arithmetic or logical operations; with qualitative variables, logical operations can be done.

Example: Disposable income for individuals = earned income + income from capital + positive transfer payments – tax – negative transfer payments

Example: Number of consumption units in a household. Swedish definition:
A household with one adult is 0.2 · 1 + 0.96 · 1 = 1.16 consumption units
A household with two adults is 0.96 · 2 = 1.92 consumption units

Number of consumption units in a household = 0.2 · (if only one adult) + 0.96 · (number of adults) + 0.76 · (number of children aged 11–17) + 0.66 · (number of children aged 4–10) + 0.56 · (number of children aged 0–3) *Logical condition underlined = 1 if true, 0 otherwise*

3. Variables derived by adjoining

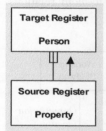

This involves creating a derived variable in a register using variables from another register. The objects in the first register can be linked to objects in the second register in a *one-to-one* relationship or a *one-to-many* relationship. This means that every object in the source register can be linked to one or many objects in the target register. Using this relationship, variables in the source register can be adjoined to the objects in the target register.

Example: In a register on individuals, the geographical coordinates of the dwelling can be adjoined to each person. Registers on individuals should contain the identity of the property or dwelling where the person is registered. The property identity is the link to the Real Estate Register. The property's coordinates are transferred over from the Real Estate Register to the relevant register on individuals. Here, properties and individuals are linked in a *one-to-many* relationship, where one property is linked to one or many individuals.

4. Variables derived by aggregation

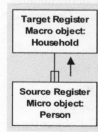

This involves creating a derived variable in a register using variables from another register. The objects in the source register can be linked to the objects in the target register using a *many-to-one* relationship. One or many objects in the source register can be linked to one object in the target register. It is possible to aggregate values, in a way that is relevant for the survey, for the *micro objects* in the source register that is linked to the respective *macro object* in the target register.

Example: Household income is an aggregated variable formed by adding the values of the variable income of individuals for all individuals in a certain household. Household is the macro object and person is the micro object.

Example: How can information from a register on individuals be combined with information from a register on enterprises? For enterprises, a derived variable is formed, *share of persons with higher education.* This variable for the macro object enterprise is formed by calculating the share of persons with higher education among all the employees (the micro objects) at the enterprise.

Mismatch will give rise to missing values in the derived variables for both adjoined variables and aggregated variables. In a well-functioning register system with a low number of mismatches, there are good possibilities for forming statistically interesting adjoined and aggregated variables.

3.3.4   Adjoining, aggregation and the structure of the register system

When we create adjoined or aggregated variables, we are matching registers containing different objects that have a certain kind of relation to one another. Via this relation, variables for one kind of objects are transformed into variables for another kind of objects. When the relation is *one-to-one*, this transformation is simple, when the relation is *one-to-many* or *many-to-one* the transformation consists of adjoining or aggregation.

The relation *many-to-many* is complicated and should be avoided. In Chapter 2, the structure of the register system is discussed. The first conceptual model of the system is shown in Chart 2.2, where relations between different object types are illustrated. In the second conceptual model in Chart 2.3 we introduce the Activity Register as the fourth base register. The contents of Charts 2.2 and 2.3 are shown in Chart 3.13 below, where we have indicated by forks if relations are *many-to-many*, *one-to-many* or *many-to-one*. By introducing the Activity Register with working and studying in the model, *many-to-many* relations are avoided and matching processes become easier to understand.

Chart 3.13 Avoid many-to-many relations!

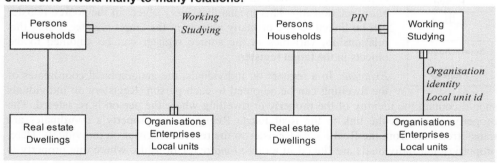

Integrating registers with different object types

Adjoining and aggregation is explained by the following example where we start with three registers before any matching or creation of derived variables has been done. One person can have many jobs and one local unit can have many employees.

Wage sum is used as name for three different variables:

– Wage sum for *job*, gross annual pay for the job that one person has at one job.

– Wage sum for *person*, aggregated gross annual pay for all jobs of one person.

– Wage sum for *local unit*, aggregate gross annual pay for all jobs at one local unit.

The example illustrates how data from three registers with different kinds of object types are integrated. A number of derived variables are created by adjoining and aggregation.

Chart 3.14A The relations between persons, activities and local units

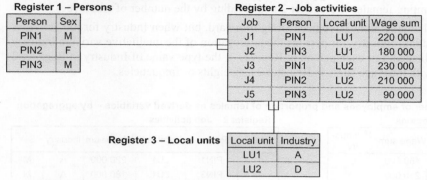

In *the first step* illustrated in the chart below, wage sums for persons and local units are derived by aggregation of job's wage sums. Data for jobs are aggregated into one value for each person or local unit. In Chart 3.14B below there are three different variables 'wage sum' defined for three different object types – persons, job activities and local units.

Chart 3.14B Wage sums for persons and local units created by aggregation

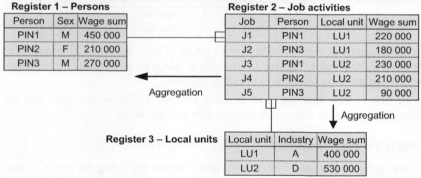

In a second step we can create derived variables for the job activities in Register 2 by adjoining variable values from Register 1 and 3. This is illustrated in Chart 3.14C below.

Chart 3.14C Industry and sex as derived variables for jobs created by adjoining

Register 1 – Persons

Person	Sex	Wage sum
PIN1	M	450 000
PIN2	F	210 000
PIN3	M	270 000

Register 2 – Job activities

Job	Person	Local unit	Wage sum	Industry	Sex
J1	PIN1	LU1	220 000	A	M
J2	PIN3	LU1	180 000	A	M
J3	PIN1	LU2	230 000	D	M
J4	PIN2	LU2	210 000	D	F
J5	PIN3	LU2	90 000	D	M

Adjoining

Adjoining

Register 3 – Local units

Local unit	Industry	Wage sum
LU1	A	400 000
LU2	D	530 000

In a third step illustrated in Chart 3.14D below, we can create more derived variables by aggregation of Industry and sex in Register 2. The variable *number of employees* in Register 3 has been created by counting jobs in Register 2 and the proportion of females has been created by counting females in Register 2 and dividing by the number of employees.

Up to now, all aggregations have been straightforward, but when Industry for job activities are aggregated into Industry for persons the aggregation of the qualitative variable Industry is more complicated. For each person in Register 1, the type value of Industry in Register 2 is computed with the wage sums in Register 2 as weights or frequencies.

Chart 3.14D
Industry, number of employees and proportion of females as derived variables – by aggregation

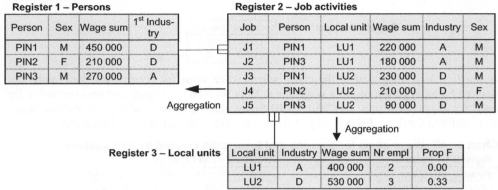

Register 1 – Persons

Person	Sex	Wage sum	1st Industry
PIN1	M	450 000	D
PIN2	F	210 000	D
PIN3	M	270 000	A

Register 2 – Job activities

Job	Person	Local unit	Wage sum	Industry	Sex
J1	PIN1	LU1	220 000	A	M
J2	PIN3	LU1	180 000	A	M
J3	PIN1	LU2	230 000	D	M
J4	PIN2	LU2	210 000	D	F
J5	PIN3	LU2	90 000	D	M

Aggregation

Register 3 – Local units

Local unit	Industry	Wage sum	Nr empl	Prop F
LU1	A	400 000	2	0.00
LU2	D	530 000	3	0.33

In the third step, inconsistencies were created in this system of three registers. Total number of employees is *three* in Register 1, but *five* in Register 3. Wage sums by Industry, in Register 1, differ from wage sums by Industry in the other registers.

The conclusion of this is that matching and creating derived variables can give rise to difficult methodological issues. In Chapter 9, these issues are discussed.

3.3.5   Variables in the register system

In this section, we describe the register system's variables, firstly regarding the origin of the variables and secondly regarding their function in the system.

Variables with different origins
A variable can be of local origin or a variable can be imported. A variable can be a primary variable or a derived variable. By combining these two concepts we get four kinds of variables with different origins:

1. Taken directly from an administrative register; certain processing may have been carried out (the format can have changed and variable values can have changed after editing). These variables are called *local primary variables*.

2. A *locally derived variable* is a derived variable created in the relevant register. All kinds of derived variables as adjoined and aggregated variables are included here.

3. Taken from the primary variables in another statistical register with identical objects, these variables are called *imported primary variables*.

4. Taken from the derived variables in another statistical register with identical objects, these variables are called *imported derived variables*.

These concepts are useful when the register is documented. For a specific register, only local variables need to be documented. Documentation of imported variables should be simply transferred when the variables are imported. Certain local variables that are important for many registers can be designated *standardised variables,* with extra high demands on documentation and quality.

Chart 3.15 Simplified chart of the Income & Taxation Register for individuals

Local primary variables			Locally derived variables			Imported variables	
PIN	Income1	.	Income200	Derived Inc1	.	Disp Income	Highest educ	ResCounty
1	10 923	.	5 223	25 766	.	197 870	5	01
2	2 344	.	0	2 344	.	23 411	0	05
.	.	.	.	.	.	.	.	.
N	73 678	.	4 311	112 973	.	213 560	6	17

The local primary variables in the data matrix above are based on the administrative data submitted by the National Tax Board to the Income and Taxation Register. These can be used to form locally derived variables. Two further variables have been imported: *Highest education* is a derived variable from the Education Register and *ResCounty* is a primary variable from the Population Register that identifies an individual's current county of residence. Highest education in Chart 3.15 is originally a multi-valued variable, as a person can have e.g. two doctorate degrees. Via a transformation rule this variable becomes a single-valued variable with the definition 'most recently taken degree'.

Variables with different functions in the system
We differentiate between six types of variables, each with a different role in the register system:

1. *Identifying variables* such as identity number, etc. are used to precisely identify objects. The corresponding IT term is *primary key.* An identifying variable should, if possible, be completely stable, i.e. it should have the same value during the whole lifetime of the object. Identifying variables are used when registers containing the same object type are matched to find hits between identical objects. Variables such as name, address, etc can also be used, but they are more troublesome variables for matching. Therefore, it is preferable to use identity number when processing registers.

2. *Communication variables* such as name, address and telephone number are used when the statistical office needs to contact an object regarding a questionnaire or an interview.

3. *Reference variables (foreign keys)* are used to describe relationships between different objects. When matching registers that contain data on different objects, reference variables produce hits between related objects.

4. *Time references* are variables that give a point in time for an event that affects objects or updates in the register. These variables are used when different register versions are created, such as the population at a specific point in time, and to describe the flow of demographic events during a given period of time.

5. *Technical variables* or variables for internal *register administration.* These variables often show the source or have comments on individual items or measurements. E.g., the source for an enterprise's industrial classification code could be the Patent and Registration Office or the National Tax Board. They can also be used to show which values have been imputed, show correction codes or error codes. Variables with weights are used for estimation.

6. The actual *statistical variables* are used when the data matrix is analysed and described. Certain variables, *spanning variables*, are used to define the cells in statistical tables, for every cell in a table; descriptive measures are calculated for other statistical variables, *response variables*.

A statistical variable can sometimes be a spanning variable in one context and a response variable in another context. When, for example, average salary is calculated for different sexes, sex is the spanning variable and salary is the response variable. If the share of women is calculated for different occupations, occupation is the spanning variable and sex is the response variable. Base registers should contain spanning variables that are important for many users; this will promote consistency.

Variables should be documented in different ways, depending on which of these six functions they have in a register. It is important that the actual statistical variables are well documented.

Variables used for matching

Here we should differentiate between two types of matches:

- The purpose of matching is to find *identical objects* in different registers or database tables. When matching, one or several identifying variables that exist in the relevant registers are used as linkage variables.
  Example: Two registers on individuals are matched; the linkage variable personal identification number exists in both.

- The purpose of matching is to find objects that have a *certain type of relationship* to one another. These objects can be found within the same database table or in different registers concerning different object types. When matching, a reference variable from the first register and a corresponding variable from the second register are used as linkage variables.
  Example: A register on individuals contains the identifying variable personal identification number but also a reference variable giving the personal identification number of that person's spouse. Two copies of this register are matched using the personal identification number as the primary key in the first copy and the reference variable as foreign key in the other copy.
  Example: A register on individuals with personal identification number as primary key can be matched against an activity register in which the gainful activity is identified by three variables: personal identification number, local unit number and organisation number. Personal identification number in the activity register is the foreign key when matching.

When two registers A and B, are matched, three new object sets are created: matching objects AB, mismatch from A and mismatch from B. All these three object sets should be saved and documented. When three registers A, B and C are matched, seven new object sets are created that need to be documented: matching objects ABC, matching in two registers (AB, AC or BC) and mismatches from A, B and C.

Chart 3.16 Combining object sets – interpreting mismatch

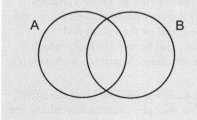 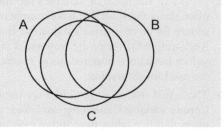

When different registers are combined by matching, as a rule different kinds of mismatch arise, and it must be decided how different categories of objects should be handled. What categories should be included in the new register, and also the causes behind mismatch should be investigated.

Matches are made using *linkage variables,* which is a summarised term for identifying variables and reference variables. A *link* between two registers consists of one or several linkage variables. The statistical term *link* corresponds to the IT term *key.* An identifying variable can be a *primary key* and a reference variable can be a *foreign key.*

Register types and variable types

Summarising the discussions in Chapters 2 and 3, different types of registers in the register system can be compared. There are differences between registers with regard to the types of variables that are important. Furthermore, different types of registers can differ with regard to the type of work they are intended for and the register's role in the register system.

Chart 3.17 A register's primary role in the system

Type of register	Types of variables	Role and responsibility
Base register	*Local primary variables:* Identifying variables Communication variables Reference variables Time references	Receive administrative data Create object sets Define objects Create some basic spanning variables Produce demographic statistics
Primary register	*Local primary variables:* Identifying variables Statistical variables	Receive administrative data Create the actual statistical variables
Integrated register	*Imported variables:* Identifying variables Statistical variables Locally derived variables, adjoined and aggregated variables	Create new information without data collection Compile information from different fields Compile information from different time periods

CHAPTER 4

Sample Surveys and Registers

A statistical office will carry out both sample surveys and register-based surveys. These two kinds of surveys can benefit from each other and can be combined in different ways. We give here also a comparison of the methodologies used in sample surveys, censuses and register-based surveys.

4.1 HOW CAN SAMPLE SURVEYS BENEFIT BY THE REGISTER SYSTEM?

The existence of a well-developed register system has important consequences for the possibilities to do sample surveys. The register system can be used in the following ways:

- When selecting the sample, the appropriate base register is used as a sampling frame and register variables are used to stratify the population.

- Measurements can be made easier by eliminating the need for questions on data that already exists in the registers.

- During the estimation phase, register variables can be used as auxiliary information to increase precision and compensate for nonresponse.

Chart 4.1 The register system and the system of sample surveys

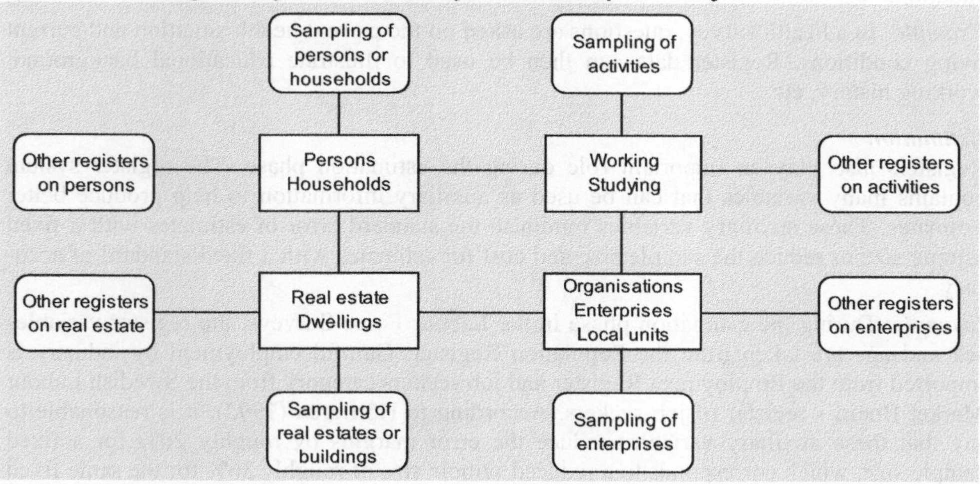

Register-based Statistics – Administrative Data for Statistical Purposes A. Wallgren and B. Wallgren
© 2007 John Wiley & Sons, Ltd

Sampling

The base registers are used as sampling frames and register variables are used to produce stratified samples. Within strata, simple random sampling is used as a rule. The base registers contain identities and addresses to objects in the frame population. By these identities the objects in each sample can be linked to other registers in the system, and register variables can be imported to the data matrix of a sample. In Chart 4.1 above, the structure of the entire statistical system is illustrated, where the sample surveys are added to the system of statistical registers.

Example: For Statistics Sweden's Labour Force Survey, the Population Register is used to select a sample among the population aged 16–64, where the sample is stratified by region, sex and citizenship according to the Population Register, and also stratified into gainfully employed/not gainfully employed according to the Employment Register. Each respondent's level of education is imported from the Education Register and the industrial code, of the local unit where the respondent works, is imported from the Business Register. Regional codes are imported from the Geographical database.

Data collection – measurements

As it is not necessary to ask for data that already exists in Statistics Sweden's registers, the burden on respondents is reduced, partly by cutting down interview time and partly by saving the interviewers from having to ask sensitive and difficult questions on, for example, income, age and education.

Example: The Income Distribution Survey primarily uses income variables from the Income Register. The selected persons only need to answer questions on household composition, occupation, whether they work full- or part-time and on certain kinds of income not recorded in the register.

Another way to combine a sample survey with register data is to survey the current situation with a questionnaire or interview and use the register data to survey the historical situation.

Example: In a health survey, questions are asked on the current health situation and current living conditions. Register data can then be used to illustrate educational background, working history, etc.

Estimation

Registers also play an important role during the estimation phase. The register system contains many variables that can be used as auxiliary information to help produce better estimates. These auxiliary variables minimise the standard error of estimates with a fixed sample size or reduce the sample size and cost for estimates with a fixed standard of accuracy.

Example: During the estimation phase in the Labour Force Surveys, the register variables sex and age are taken from the Population Register. Gainful employment by industry is imported from the Employment Register and job seekers category from the Swedish Labour Market Board's register of job seekers. According to Hörngren (1992), it is reasonable to say that these auxiliary variables reduce the error margins by roughly 20% for a fixed sample size, which corresponds to a reduced sample size of roughly 36% for the same fixed accuracy standard.

Nonresponse in sample surveys is a significant problem and the size of nonresponse is increasing. Register data can be used to *reduce the size* of the nonresponse as questions about certain sensitive variables such as income, education and age can be avoided. Questionnaires also become shorter if register variables are used instead of questions. This can also help reduce the size of the nonresponse.

Register variables can also be used to minimise the effects of *nonresponse error*. It is therefore an advantage to have access to many variables so that it is possible to choose the most suitable nonresponse adjustment method for the survey in question. Särndal and Lundström (2005), give a comprehensive description of how register variables can be used for calibration of sampling weights to reduce sampling errors, nonresponse errors and frame errors.

4.2   COMBINING REGISTER-BASED SURVEYS AND SAMPLE SURVEYS

Register-based surveys and sample surveys or other kinds of data collection can be combined in many ways. We give here a short description of some cases where the combination of these two methods is important.

1. Defining a precise target population
A register-based survey and a sample survey can complement each other in terms of content. Register-based statistics give basic data on the differences between different categories and on changes over time. It is then possible to carry out a sample survey, which gives a more detailed description of the reasons for these patterns. By using registers, the sample can be limited to sub-groups that are specifically of interest.

Example: By using the Value Added Tax Register, it is possible to select a sample of enterprises (legal units) that have had a lower turnover over the past year. These selected enterprises can then be interviewed about the reasons for the decline and about what they predict for the future.

Example: Using the Activity Register, it is possible to select a sample of persons who have changed employer during the past two years and interview them on the reasons for these changes.

2. Sample surveys can give indications on register quality
The quality of the base registers is of vital importance for the whole register system. By analysing sample surveys done for other purposes, it is possible to get indications on overcoverage. A business survey can give an estimate of the proportion of inactive enterprises in the Business Register. For surveys of persons and households, the nonresponse in telephone interviews or returns of postal questionnaires can be analysed to get indications of overcoverage in the Population Register.

The quality of important classification variables in, for instance, the Business Register should be monitored by special sample surveys where branch of industry and other variables are checked. All ordinary sample surveys with variables similar to variables in statistical registers should be used to get indications on measurement problems. The causes of these errors can be found in both the sample survey and in the register. This is also discussed in Section 5.4.7.

3. Register maintenance surveys should be used to improve register quality
Certain categories of objects in a base register can be imperfect, important register variables can be missing or can be suspected to be of low quality. Mail questionnaires or telephone interviews to such categories will improve the quality of the register and should be used as regular methods to maintain high quality of a base register.

4. An administrative register can be complemented with special data collection
An administrative source with statistically important variables can be incomplete due to undercoverage. Small enterprises, for instance, can be missing in some administrative systems. If such an administrative register is compared with the appropriate base register it may be that there is a category of objects in the base register, which is not found in the administrative register in question. The population should in this case be divided into two strata: one stratum that is found in the administrative source, and a second stratum that is in the base register, but not in the administrative source. The first stratum is studied by a register-based survey, and the second stratum is studied by a sample survey or census. This method is discussed by Selander et al. (1998).

5. Sample surveys can be used when creating derived variables in registers
Statistical variables in sample surveys can be compared with administrative variables in registers. It is possible to build measurement models that can be used to find out how the administrative variables should be used when creating derived variables for statistical purposes. This is discussed in Section 6.2.3.

6. Small area estimation
"Small area estimation of variables studied in social surveys is a growing need for government, principally for the establishment of better directed resource allocation for problems of bad health, bad housing conditions, unemployment and low pay." Heady et al. (2003) introduce their report on small area estimation in this way, indicating the important demand of statistics describing small geographical areas.

Heady et al. develop methods for small area estimation for England and Wales. In the Scandinavian countries, the same demand is met by developing register-based surveys, but for countries where the statistical offices don't have access to administrative registers with rich content that can be used for statistical purposes, their experiences with small area estimation is of great interest.

Their report describes also the long-term efforts, which are necessary for building up the competence regarding this kind of statistical model building. The methods they have developed are based on a combined use of data from sample surveys and administrative data or data from the previous census. The micro data from the sample is increased with aggregated data from administrative registers or aggregated census data. The observation for each person in the sample then contains data from the interview or questionnaire used in the sample survey and covariates in the form of aggregated register or census data for that person's small geographical area. With multilevel regression models each sample survey variable of interest is explained by the covariates. These estimated models together with aggregated administrative or census data for all small geographical areas are used to produce estimates for all areas.

7. Virtual censuses
A traditional population and housing census is based entirely on data collection. A register-based census is completely based on administrative registers, but requires full access to administrative sources with all information necessary.

The term 'virtual census' refers to a census based on a combination of administrative registers and sample surveys. In Statistics Netherlands (2004), the Dutch virtual census 2001 is described. A virtual census requires some administrative sources, such as a population register, but when the administrative sources lack information this is supplemented by sample surveys.

The Dutch virtual census of 2001 was completely based on already existing sources. The costs saved by this approach were substantial – the cost for a traditional census would have been about 300 million euros, while the cost for the virtual census was only about 3 million. There were other reasons for developing a virtual census, the willingness to participate in a traditional census had fallen, but a virtual census is more socially acceptable. The participation in a traditional census would be limited and selective, resulting in difficult problems regarding how to correct for nonresponse errors.

An important methodological issue is how data from different administrative registers, and from different sample surveys, can be combined to produce consistent estimates resulting in the census tables to be published. Houbiers et al. (2003) describe the method of repeated weighting, which is used to adjust the estimates based on the sample surveys, so that all estimates published will be consistent.

Small area estimation methods can also be used in a virtual census; this is mentioned as an interesting possibility for the future development of the Dutch virtual censuses. When more administrative sources become available, a virtual census can be replaced by a register-based census.

4.3   COMPARING SAMPLE SURVEYS AND REGISTER-BASED SURVEYS

Planning a register-based survey is completely different from planning a sample survey. The first step with a *sample survey* is to decide on the population and which parameters should be estimated for which domains of interest. This determines the character of the survey with regard to sample design and estimation. The population and parameters are defined first, followed by the collection of data. In general, *one* survey is considered at one occasion with a limited number of parameters and domains of interest. When working with sample surveys, the concept pairs *population – sample* and *parameter – estimator* are often used.

With a *register-based survey*, the starting point is very different, as data have already been collected and are available in different registers, where they are not adapted for any specific survey. Using the available registers, the objects and variables that are relevant for the survey in question are selected. It is sometimes the case that it is necessary to form new variables from the existing registers – and sometimes even new objects (statistical units). But, in the beginning, the data already exist and then the population is defined. The plan of tables is then decided without thinking in terms such as parameter estimation and domains of interest. Sampling error does not limit the possibilities for choosing domains for the analysis and the reporting of results. When working with register-based statistics, the distinction between *population – sample* has no relevance and it is not usual to use the terms *parameter – estimator*.

When designing a statistical register and a register system, it is desirable to make them flexible so that they are as widely applicable as possible. Therefore, an important part of register-statistical methodology work is to structure and improve the whole, i.e. to find the best possible design for the entire register system. This includes long-term work to monitor, and influence the access to administrative data for statistical purposes.

So although much of statistical methodology can be considered the same for sample surveys and for register-based surveys, such as problems with errors (excluding sampling error) and the work with analysis and presentation, the way of thinking is different, as sampling error and design problems are so crucial for sample surveys. For register-based statistics, *system-based thinking* is fundamental. To improve quality, it is not sufficient to look at one register at a time but the system should be seen in its entirety. Special attention should be paid to the quality of the base registers and the identifying variables that act as links between the different registers.

There are different preconditions for *the editing process* for sample surveys and for register-based surveys. If unreasonable values are detected when checking a sample survey, it is possible to re-contact the relevant respondent. Alternatively, it is possible to repeat a question during an interview if a response seems strange. These solutions are not possible with register-based surveys. However, questions can be put to the register-providing authority on what is considered a reasonable value and how errors can occur. It is important that persons at the statistical office know which variables the authority has checked and corrected.

Register-based surveys have editing problems in other situations as well. When receiving large registers, it can be necessary to check and correct technical errors. For example, data that has been scanned can still have values in the wrong positions. In certain cases, data are not collected centrally but by different regional authorities and then, at the statistical office, it can be investigated whether there are structural differences between the different authorities regarding time delays, response patterns, etc.

The problem of *nonresponse* is also of a different character for sample surveys and register-based surveys. Missing values also exist in register-based surveys, but persons who could not be contacted or refused to respond do not cause the same problem as in sample surveys or censuses, and it is not necessary to send out reminders or decide when to bring an end to the data collection. In sample surveys and censuses, estimates are revised to counteract the effects of nonresponse. This does not always occur with register-based statistics.

Measurement errors exist both in sample surveys and register-based surveys. For sample surveys, we focus on minimising measurement error by testing and improving the questionnaires. We cannot, however, work in the same way with register-based statistics, as it is the administrative authorities that design the questionnaires. However, a national statistical office should have a role in influencing the different authorities so that the questionnaires and instructions can be improved.

The *presentation* of survey results has somewhat different prerequisites. We cannot present results from sample surveys for groups that are too small, the limits being set by the sampling error. The tables presented can therefore not be broken down by many variables at the same time and cannot have too many cells. However, with register-based statistics, it is possible to produce large detailed tables that are broken down in many ways. Such tables are often difficult to interpret, and place great demands on the method of presentation. There are also limits for how detailed the tables can be due to *risks for exposure* for the separate individuals and enterprises.

These examples illustrate how sample surveys and register-based surveys often have different types of methodology problems. Methodology development in the areas of editing, nonresponse and measurement error has, until today, principally concerned sample surveys.

To avoid mixing together the different concepts, it is important to clearly distinguish between the three types of survey that exist at a statistical office. We here compare the similarities and differences in the conditions for sample surveys, censuses and register-based surveys. The latter, which will be discussed in the next chapters, has been shaded in Chart 4.2 below.

Both data collected for censuses and registers based on administrative data can be included in the register system if the data contain identifying variables so that the data can be linked with other registers. However, sample surveys are not parts of the register system. Sample surveys can use registers in the system, but as sample survey data can't be used by registers in the system they are not included in the register system.

Chart 4.2 Similarities and differences between the different types of survey

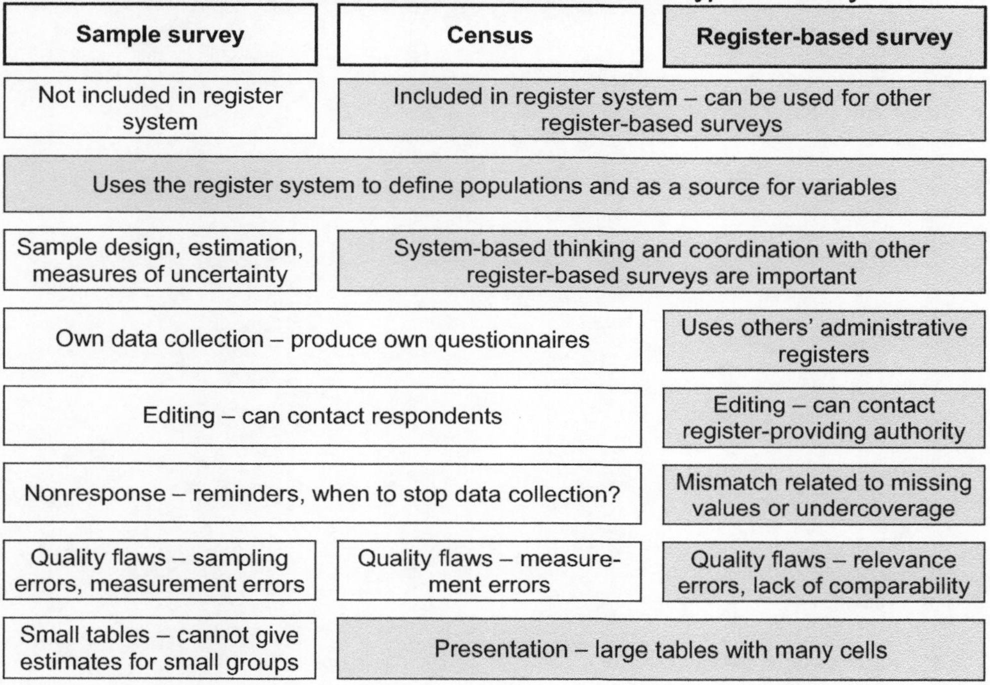

Sample survey	Census	Register-based survey
Not included in register system	Included in register system – can be used for other register-based surveys	
Uses the register system to define populations and as a source for variables		
Sample design, estimation, measures of uncertainty	System-based thinking and coordination with other register-based surveys are important	
Own data collection – produce own questionnaires		Uses others' administrative registers
Editing – can contact respondents		Editing – can contact register-providing authority
Nonresponse – reminders, when to stop data collection?		Mismatch related to missing values or undercoverage
Quality flaws – sampling errors, measurement errors	Quality flaws – measurement errors	Quality flaws – relevance errors, lack of comparability
Small tables – cannot give estimates for small groups	Presentation – large tables with many cells	

CHAPTER 5

How to create a Register
– The Population

The need for methodology is as great for register-based surveys as it is for sample surveys. When we started our work with register-based statistics, we found that many units at Statistics Sweden had carried out advanced register processing for a long time, which has led to the development of significant knowledge about how register-based statistics should be produced. Methods were obviously used but no generally formulated methodology existed as a guideline. In this chapter and the next, we formulate some general concepts and principles that can serve as a first step towards a common register-statistical methodology.

The traditional survey methodology provides an answer to the question, *'how should sample surveys and censuses be carried out?'* In the same way, register-statistical methodology should give an answer to the question, *'how should a register-based survey be carried out?'*

However, because any one statistical register is not only designed for *one* specific survey but should, either alone or in combination with other registers, be used for several different surveys, the thinking behind the methodology must be more flexible. The register-statistical methodology should therefore also provide an answer to the question *'how should a statistical register be created so that it can be used for a certain specific survey but could also contribute to the possibility of carrying out other surveys within the register system?'* These are questions which should be discussed both when considering a new register and also when developing existing registers.

5.1   HOW SHOULD REGISTER-BASED SURVEYS BE STRUCTURED?

Work with register-statistical methodology is carried out on three levels:

- *Work at system level.* This work involves the structuring and coordination of a large number of registers into one functioning system of statistical registers. The system as a whole should be developed to make it possible to produce new and better statistics.
- *Work to create a statistical register within the existing system.* This work involves responsibility for the methods used when a statistical register is created. How should administrative data be used to create a primary register? How should a base register be constructed? How can the register system be used to build an integrated register?
  The work aims to create the data matrix or matrices that the register consists of. These types of questions are discussed in this and the next chapter.

Register-based Statistics – Administrative Data for Statistical Purposes   A. Wallgren and B. Wallgren
© 2007 John Wiley & Sons, Ltd

– *Work with a survey based on an existing statistical register.* Based on the research objectives, the work aims to analyse the register's data matrices using different calculations, aggregations and methods of estimation. How should calculations be carried out to take into account different methodology issues, such as coverage problems, missing values and level shifts in time series? How should registers with multi-valued variables be processed? Register processing results in a number of statistical tables for a specific survey. This type of question is dealt with in Chapters 8 and 9.

Administrative data must be processed to meet the requirements of the statistical purposes. Using all the available sources that have a connection with the statistical register to be created, the register's object set, objects and variable content are defined in the best way possible. What register processing should be carried out? How should this be done? In this chapter and the next, we describe this processing from a statistical perspective and in Chapter 11 we give a brief description of the processing from an IT perspective.

The procedure of creating a statistical register
How should a statistical register be created, which will partly be used for certain specific surveys and which will partly contribute to the possibilities of carrying out other surveys within the register system? The work with sample surveys is usually divided up into different stages and, in the same way, the work to create a statistical register can be divided into different phases:

1. Determining the research objectives: What statistical requirements and aims are to be fulfilled with the register? This is discussed in Section 5.2.

2. The inventory phase: What sources are available when a new register is to be created? What administrative sources are available and what existing statistical registers in the system can be used? This is discussed in Section 5.3.

3. The administrative sources and existing statistical registers shall be integrated into a new register. This integration process must be planned. This planning phase can be divided into three parts:

   a. Definition of the register's object set. This is discussed in Section 5.4.

   b. Definition of the register's objects (statistical units). This is discussed in 5.5.

   c. Definition of the register's variable content. This is discussed in 6.1 and 6.2.

4. Contacts with data suppliers and the receipt of administrative data: Checks and editing of the administrative data received are discussed in Section 6.3.

5. The integration phase where the different sources are integrated into a new statistical register can be divided into three parts:

   a. How should the existing sources be integrated so that the register will contain the required object set? The administrative data is checked and edited so that the object set is the one required. Different sources are matched and objects are selected. Time references are processed to create the object set for the point in time or period in question. This is discussed in Section 5.4.

   b. What processing should be carried out to check and correct object definitions? Administrative data is checked and edited so that the object definitions match those required. Derived objects are formed in the new register. This is discussed in Section 5.5.

c. What processing should be carried out to create the variables in question? The variables in the administrative sources are checked and edited. Steps are taken if variable values are missing. Different sources are matched; variables are selected and imported to the new register. Derived variables are formed in the new register. This is discussed in Chapter 6.

What processing should be carried out? The register to be created is a base register, a primary statistical register or an integrated register. The demands on processing can be different for these three types of registers.

Work with a register-based survey should be structured so that the persons working with the survey are aware of the three processes that are ongoing at the same time. The first process consists of work to create the statistical register and is described in points 1-5 above. In parallel with this, the register and processing should be *quality assured* and *documented.*

Quality assurance

Work to create statistical registers should be quality assured. The register's quality should be reviewed and described using various quality indicators. Documentation is also an important part of quality assurance. Incorrect and uncritical use of administrative data can be prevented by the existence of metadata, which give information on possible comparability problems. These issues are discussed in Chapter 10.

Documentation

As a statistical register should be able to be used by many users who are utilising the register system, all the registers should be documented in a way so that everyone can use and understand the documentation. Metadata has a very significant role in the work with register-based statistics. When linking and matching different registers, it is necessary to know the definitions and any comparability problems. It is also important that the processing methods are documented to facilitate the development of methodology and exchanges of experience. Chapter 11 discusses metadata and documentation.

Chart 5.1 The three parallel processes with a register-based survey

Create and use a register	Quality assurance (Chapter 10)	Documentation (Chapter 11)

Creating a register (Chapters 5 and 6)
 	
1. Determining the research objectives
2. Inventory phase
3. Planning phase
4. Receiving data
5. Integration phase:
  – create population
  – create objects
  – create variables
 	
Calculations and tabulations (Chapters 7, 8 and 9)
 	
Selection of estimation methods for carrying out calculations and compiling tables
	Contacts with data suppliers
	
	Checking on receipt of data
	
	Causes and extent of missing values
	
	Causes and extent of mismatches
	
	Evaluating quality of objects and variables
	
	Questionnaires for register maintenance
	
	Investigating inconsistency between different sources
	
	Documentation
		When the microdata is collected from different source registers, metadata for the sources should also be collected:
		
		Register populations for every source
		
		Definitions of imported variables
		
		What checks and processing have been carried out in the different sources?
		
		What is known about the quality of the different sources?
		
		The new register should then be documented

5.2   DETERMINING THE RESEARCH OBJECTIVES

Which surveys need to be done, what questions need to be answered by these surveys? The work with defining the research objectives should give the answers to these questions. This phase is of fundamental importance for the quality and relevance of the survey. However, it is often overlooked; a common mistake is to start working with collecting data too early so that the survey not will give the required answers. Before we spend money on expensive data collecting, we should always try to use available data. In a register-based survey we can work with defining the research objectives and at the same time see what data exist that can be of interest.

When the work with the definition of the research objectives is complete, the target population and the important statistical variables should have been defined. We will later in this chapter discuss the definition of the target population and in Chapter 6 we discuss the definition of statistical variables. These definitions should be both as adequate and functional as possible.

Different users need different registers
We will use Chart 2.11 from Chapter 2 for a discussion of how new register-based surveys have been created at Statistics Sweden to meet the needs of different users.

That chart describes the existing registers in the system's population section. Within the system's other sections, and when discussing a completely new register, the same reasoning can be used. We have so far regarded Chart 2.11 as a part of a register system. But for every register, there are specific questions that have motivated the register. Throughout Statistics Sweden, knowledge is built up on the needs of the different users. The staff must, at the same time, be involved in the subject field, be familiar with the possibilities of the register system and know which statistical methodology could be of use.

The chart therefore also shows the structure of a survey area (statistics on individuals) and its parts.

1. The base register:	Demographic statistics: Population, births/deaths, migration

The base register lays the groundwork for demographic statistics, which aim to describe the structure of the population and how this changes over time. These demographic questions make up their own category of research objectives with a well-established methodology.

2. Registers directly based on administrative data:	Income & Taxation Register Privately owned Vehicles Patient Register Cancer Register Cause of Death Register

A number of important administrative sources have for a long time been used as the basis for statistics. Education and income are areas where the statistical registers have been built up successively in cooperation between various researchers and other users. Medical data from registers have for a long time been used for research.

3. Integrated registers for official statistics:	Register-based 'census' Employment Register Education Register

Some of the more creative employees at Statistics Sweden, with good contacts with different users, see new possibilities of using existing register data. When they succeed in bringing together the new needs with the system's possibilities, a new integrated register is created.

4. Integrated registers for research:	Multi-generation Register Fertility Register Longitudinal Income Register Longitudinal Welfare Register Education & labour market transition

Researchers can have access to the existing registers at Statistics Sweden for their own analyses. Statistics Sweden can also create registers specifically for the researchers' needs by combining different Statistics Sweden registers. All record linkage is carried out according to the Swedish Secrecy Act and only anonymised data are released after review.

5. Micro-simulation models:	The Income & Taxation Register is the basis for a simulation model where planned changes in taxation and transfer payments can be tested. The model is used by government and research.
	The Employment Register is the basis for a regional simulation and forecasting model used by government and research

Simulation models based on data from Statistics Sweden registers can show how planned changes can affect such things as income distribution or the development in a specific region.

6. Standardised regional tables:	The Population Register, Income & Taxation Register, Privately owned Vehicles, Employment Register and Education Register are the basis for standardised regional tables used by local government and market analysts.
	The regions can be chosen by GIS technology.

Using different table packages, regional statistics can be produced for any regional breakdown. Residency profiles and Market profiles are also examples of register-based products with regional tables that can be used by enterprise customers for various market analyses.

Different scope of the research objectives
A traditional sample survey, e.g. the Labour Force Survey (LFS), reports only one set of data – the data collected in the survey. The team responsible for the LFS may describe their mission as *'We collect LFS data and we analyse and report LFS data.'*

There is a risk that those working with register-based surveys also understand their mission in the same narrow manner: *'We are responsible for administrative data from source X and we analyse and report X data.'* If a team responsible for a register-based survey understand their mission in this way, all opportunities of the register system are not taken advantage of. Instead of this narrow scope, they should use all relevant registers in the system to analyse and describe their subject field.

Example: The Register of University students
This register receives administrative data from all universities concerning: what the students want to study, what they actually study and the results of their efforts and examinations. The traditional approach is to report these sources only. However, there is more information in the register system about these university students:

1. Do they also work to earn money, and how much money do they earn?

2. From where did they come?

3. What did they do before the university?

4. What happened after they finished their university studies?

If the team responsible for University statistics instead of only reporting university studies also decides to give answers to these four questions, they can use more registers in the system.

5.3   MAKING AN INVENTORY OF DIFFERENT SOURCES

After structuring the research objectives, an inventory should be made of which different sources could be used to create the new register. There are three types of sources (source registers) that could be of interest:

–   Existing statistical registers in the register system of a statistical office. Information on what these registers contain should be clear from existing register documentation.

–   Administrative sources known within the statistical office, but not currently used in the register system. For example, there can be parts of deliveries that are currently not used at the statistical office or parts of the administrative registers that the statistical office does not request from the administrative authority. A third possibility is to carry out an *integrated data collection,* i.e. the administrative authority adds a question on their questionnaire specifically for statistical purposes. Information on these possible sources can be found in the unit at the statistical office that has contact with the respective authority.

–   Completely new administrative registers that no one at the statistical office is currently aware of. These could be administrative sources within enterprises or authorities that are still not used for statistical purposes, but that could be used. We are talking here about completely new sources, which is why it is important to build up contacts outside the statistical office.

All available sources with a connection to the problem area should be analysed when the new register is created. This is an important principle, as every new source could potentially be used to improve coverage in the new register regarding objects and variables. A new source can also mean that inconsistencies are detected which can contribute to improved quality.

A source, which contains quality flaws, could also be used together with other sources. The source alone would perhaps produce register-based statistics of low quality but, *as one part of the register system,* it could still contribute to the improvement of the system's overall coverage and quality.

5.4   DEFINING A REGISTER'S OBJECT SET

In every statistical survey, the population of the survey needs to be defined. In this section we discuss the definition of the object set and then deal with the definition of objects or statistical units in the next section.

When a register could be used for several surveys, it should be possible to define different populations with the register. To meet these demands, it should be possible to make different object selections from the register so that the different sub-sets suit the different surveys. Another possibility is to create different versions of the register for different applications.

We here reserve the concept *population* to refer to an object set that belongs to a specific survey. When we describe a register without referring to a specific survey, we use the concept *object set*.

However, every statistical register is created for one or several principal uses or surveys. It is therefore common that the register's object set agrees with the main survey's population. When creating a new register for a specific survey, the new register's population should be defined. Every source register has its own object set that will be included either completely or partially in the new register.

5.4.1 Defining a population

Every survey begins with a set of questions that are formulated in theoretical or general terms. The theoretical concepts in the set of questions must then be operationalised, i.e. translated into measurable concepts. When defining these measurable concepts, *what* is being surveyed is determined. A population should be defined in the following way:

> The population definition should clearly show which objects are included in that population. The object type should be clearly specified. In addition, a time reference and geographic delimitation should always be included. The geographic delimitation should also specify the relation that exists between the objects or statistical units and the geographical area.

Example of a population definition:
'Permanently resident individuals in Sweden on 31/12 2004. Permanently resident refers to ...'

individuals	= object type
permanently resident in	= relation
Sweden	= geographical area
on 31/12 2002	= point in time

Sample survey theory and guidelines for quality concepts and quality declarations by statistical offices usually contain three concepts related to populations:

– *Population of interest* refers to the population in the theoretical question at hand.

– *Target population* refers to the operationalised population, the theoretical population of interest, which has been translated into a concrete and examinable population, i.e. the population that it is the *target* for the survey.

– *Frame population* refers to the object set that the *frame* actually gives rise to. These concepts can be found in the theory of sample surveys and censuses with their own data collection.

For register-based surveys, we only use the two former concepts. Because the sampling frame does not exist with register-based surveys, the third concept *frame population* must be replaced:

– *Register population* refers to the object set in the register that has been created for the survey in question, i.e. the population that is *actually* being surveyed.

There are important differences between a frame population and a register population. A frame population is defined *before* the data collection, while a register population is created *after* the reference period when all administrative data has been received. A frame can consist of other objects than those to be surveyed, i.e. a map or an address register for a survey on households.

Example: The structure of the population using the Population Register
Here, we want to survey the size and structure of the Swedish population on December 31 of a specific year. The *population of interest* is permanent residents in Sweden on December 31, but this vague concept needs to be further defined. It can be defined in different ways. In general, a good definition should meet the following requirements:

– it should be *adequate*, i.e. it should be in accordance with what you want to survey;

– it should be *functional,* i.e. it should be applicable in a practical sense.

When setting up a definition, it is often difficult to find the balance between what you want to survey (adequate definition) and what it is possible to survey (functional definition). In this example, the following definitions of

'permanent residents in Sweden on 31/12 of year t' could be possible*:*

(1) Persons registered by Tax Board in Sweden on 31/12 of year *t*

(2) Persons registered by Tax Board in Sweden on 31/12 of year *t*, according to data available at the end of January of year *t+1*

Those working with the Swedish Population Register usually wait until the end of January to create the register version that relates to the situation on 31/12. It is then hoped that all the changes and events affecting the population register for the previous year have been reported in. The register then created should be the version that is applied even if some notifications referring to year *t* can still be received. Therefore, the *register population* is defined using definition (2) for this survey.

However, it states in the quality declaration for the Swedish annual population statistics that the survey aims to describe the *target population* according to definition (1) given above. The difference between the target population and the register population is therefore the register's *coverage error*. Deaths and emigrations that have not been reported before the end of January cause *overcoverage*, while births and immigrations that have not been reported cause *undercoverage*.

In addition to these coverage errors, relevance error can also occur if the definition used is not adequate. The difference between the population of interest and the target population is one of the survey's *relevance errors*. There are likely between 25 000–50 000 persons registered in the population register in Sweden that do not live permanently in Sweden. It is judged that 4–8% of immigrants from outside the Nordic countries have left Sweden without reporting this. This relevance error affects statistical estimates describing the death rates, average income, etc. for immigrants from outside the Nordic countries so that the estimates become misleading.

In the example above, we can see that Statistics Sweden's population statistics use an administrative definition, the registered population, when defining the survey's target population. Administrative concepts always give definitions that are functional. It is sensible statistical practice that these administrative concepts are used to define the target population if the relevance errors are small. However, the basic rule is that the population definition should attempt to meet the demands of the statistical survey. If the administrative concepts are not sufficiently relevant or adequate, it is necessary to develop own definitions and carry out the register processing required so that the register's object set reflects the defined population as closely as possible.

Is it possible to leave the present administrative definition of Sweden's population? This is discussed within Statistics Sweden and there are ways to improve the definition of the target population:

– Include foreign students studying at Swedish universities, administrative data is available. They are not registered by the Tax Board, but by the universities.

– Exclude Swedish students studying abroad, they are registered by the Tax Board, but do not live permanently in Sweden. Administrative data is available for many of them.

– Swedish university students can in many cases be registered by the Tax Board where they lived before going to the university. Their present address is registered by the universities. This would result in a geographical reallocation of many university students; regional population statistics will be different.

This means that it is possible to leave the administrative definition of the target population and introduce a new definition, which will be more adequate.

5.4.2 Can I alter data from the National Tax Board?

Should I, being responsible for a register at a statistical office, really change administrative data that comes from another authority? The administrative authorities collect the data, and therefore have primary responsibility for the register. A person working at a statistical office with a specific product or survey may not come up with the idea that the data should be changed or supplemented to suit the statistical purposes of the product in question.

It was pointed out in Chapter 1 that administrative registers should be processed so that the objects and variables correspond to the needs of the statistics. This means that the persons at a statistical office who receive the administrative data have both the freedom and the obligation to carry out such changes so that the quality of the statistics can be improved. The persons who make these changes should be experienced, independent and have the support of a network of register statisticians. Otherwise, they perhaps will not dare make the changes.

It can be very convenient not to make any changes in the administrative data you get. Then you can say that the administrative authority is responsible for all errors. If you make changes in the data you have been given, you may feel that you are responsible for all errors. However, if you are a statistician, you are always responsible for the statistics you produce.

5.4.3 Defining a population – primary registers

How should those working with primary registers define target and register populations? Primary registers are based wholly or partially on administrative registers, and there is thus a risk that the administrative system's object sets will influence the choice of register population in an inappropriate way.

The object set in the administrative register may not completely cover the target population that is of statistical interest. The administrative object set consists solely of those objects that are included in the administrative system, and there can then be both over- and under-coverage compared with the statistically desirable target population.

Example: The administrative object set - is it suitable as the target population?
Agricultural statistics is based on the applications for subsidies that farmers in the European Union submit to the county administrative boards. These applications are registered in the

IACS system (Integrated Administration and Control System), which is used to administer the agricultural subsidies.

Chart 5.2 shows a comparison of area data in the applications for area subsidies 1995-1997 within the IACS system, and of corresponding data from the Farm Register, which was based on a census carried out by Statistics Sweden. The differences between IACS and the Farm Register are due to undercoverage in the IACS register – some farmers do not apply for subsidies even though they are actively farming.

Chart 5.2 Undercoverage in an administrative register

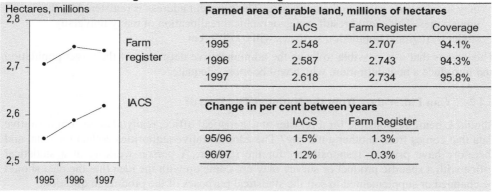

Farmed area of arable land, millions of hectares			
	IACS	Farm Register	Coverage
1995	2.548	2.707	94.1%
1996	2.587	2.743	94.3%
1997	2.618	2.734	95.8%

Change in per cent between years		
	IACS	Farm Register
95/96	1.5%	1.3%
96/97	1.2%	–0.3%

Conclusions: Although the IACS register can be considered to have good coverage, seemingly small variations in the coverage mean that the time series for farmed area of arable land is totally misleading – an actual decrease 1996–1997 appears to be a continued increase in the IACS register.

Flaws in coverage in administrative registers should not be ignored. On the contrary, target populations in these cases should be defined according to statistical requirements. Then a new register should be created containing the intended register population using the current base register, in this case by selecting objects from the Business Register. This new register can then be matched against the IACS register, making it possible to detect both any over or undercoverage in the IACS register.

Overcoverage is an indication of possible flaws in the Business Register. Undercoverage in the IACS register will appear as missing values in the new register. This nonresponse can be corrected either via nonresponse adjustment, or by adding a special survey to collect data from the part of the target population that is not included in the administrative register.

5.4.4   Defining a population – integrated registers

How should those working with integrated registers define target populations and register populations? We discuss here the basic principles for this.

Example: Register commissions with matching – selection of target population
Many register commissions involve the combined processing of several registers. A series of matchings are carried out and variables are imported from different registers. The end result is an integrated register with many variables that are interesting for the project's customers.

But how has the survey's target population been defined? It can easily be the case that the object set of the integrated register is an intersection of the matched registers' object sets.

Does this intersection represent an equivalent target population for the project? It should not be taken for granted that this is the case.

Such a register commission should also begin with the definition of the target population with regard to the problem to be studied. An appropriate object set is then selected from the relevant base register. This object set is the register population, which is then matched against the registers containing variables of interest. For those objects that receive hits, the variable values are imported to the new register. For those objects not receiving hits, item nonresponse is shown, i.e. the variable values are missing.

Chart 5.3 Object sets when matching two registers

Each ellipse represents an object set from one of two different registers.

The shaded area represents the objects that produce hits when matched, i.e. the intersection.

Conclusions: Which target population corresponds to the intersection? The intersection can never be a target population, as it would not then be possible to give a definition of this. It should not be the register population either, as nonresponse due to mismatches would risk being forgotten. It is important to be aware of the nonresponse, and also to carry out nonresponse adjustment where the scope and structure of the nonresponse is known. This is described in Chapter 8.

5.4.5 Base registers should be used when defining populations

Both those working with integrated registers and those creating primary registers using administrative registers should use the base registers when the object sets are defined. There are two reasons for this – firstly, because the object sets in the base register should be the 'best', and secondly, because register-based statistics should be consistent. Those working with primary registers and integrated registers should begin with each base register's object set and use one of the *standardised populations* that are created for general usage:

- the *end of year version* suitable for annual stock statistics (such as the population on 31/12),
- the *calendar year version* suitable for annual flow statistics (such as the population's income during a specific year),
- a *monthly or quarterly version* suitable for monthly/quarterly statistics.

General methodology:

1. Define the target population.
2. Select the intended object set from the base register, giving the register population.
3. Match against registers containing interesting variables.
4. When receiving hits: import the variable values to the register which is created.
5. When receiving mismatches: show missing values (item nonresponse).

5.4.6   Requirements for a base register

There are four requirements placed on a base register so that it can be used to define register populations:

– the base register should contain time references, i.e. all events that affect the register's objects or statistical units should be dated,
– the base register should have good coverage,
– linkage variables should be of high quality,
– spanning variables should be of high quality.

Time references

For every object, the point in time when the object was born or ceased to exist should be known. Points in time for other important events, such as moves or changed categories, should also be known.

Dates of events are important, but they are not always possible to determine. The date when the change was registered is almost as important. A base register should contain several kinds of time references: dates of event, dates of registration and dates of change.

Using these time references, it is possible to create register populations that reflect the population's status at a specific point in time or period.

Time references can have different levels of precision; events for persons are usually known so that a specific day can be given, events regarding enterprises are perhaps only known so that only the month can be given.

It is important in a base register to differentiate between statistically relevant and statistically trivial changes. The change of a postcode or national telephone code for an entire area should not appear as if the object has moved. In the same way, a change in an enterprise's legal form should not be seen as the enterprise ceasing to exist and a new enterprise being formed.

Coverage, spanning variables and identifying variables

The majority of populations for Statistics Sweden's surveys are defined using one of the base registers. This places tough demands that the object sets in the base registers have good coverage for the needs of many surveys, and that the links and variables used to select sub-populations or divide populations into domains of interest *(spanning variables)* are up-to-date and of good quality. If these variables lack quality, this can cause coverage errors in sub-populations. This will happen if data on branch of industry are missing, or if the Industry code is incorrect.

The above example in section 5.4.3 on agricultural enterprises illustrates the requirements that a base register should fulfil. In this case, it must be possible to use the Business Register to select agricultural enterprises so that the new register has good coverage of the target population. To ensure this, the Business Register needs to have good coverage of all active enterprises, and the industry code should be of good quality, so that it can be used to select the desired register population. Additionally, the Business Register contains identifying variables such as organisation number, enterprise name, address and telephone number that are important when matching against the IACS register with applications for agricultural subsidies.

To ensure good coverage in a base register, all the relevant administrative sources should be used. This is a general principle for creating a statistical register – *quality can be improved by combining many sources.*

Example: It is considered that the November version of the Swedish Business Register is of good quality, and this version is used as the sampling frame for many surveys. This good quality does not necessarily apply to small enterprises and all industry sectors. If the November version of year *t* is matched against the IACS register from the summer of year *t*, it is clear that there is significant undercoverage. By adding data from the Statements of Earnings Register (SER), Value-Added Tax Register (VAT) and Income Declarations for Enterprises (IDE) for year *t*, the coverage in the Business Register will be improved.

Chart 5.4 Results of different matches between the Business and IACS Registers

IACS register matched against	Linkage variables	Matches in IACS
Business Register, November 1995	Organisation number	75.2%
Business Register supplemented by:		
SER, VAT and IDE for 1995	Organisation number	96.6%

Conclusions: One conclusion that can be drawn from this example is that the Business Register for 1995 has poor coverage of agricultural enterprises. By supplementing the Business Register with other sources, the coverage can be significantly improved. There are four sources in this example that could be used to improve this part of Business Register. The IACS register is the most up-to-date source that becomes available roughly one year earlier than the other sources. Furthermore, the enterprises in IACS are active agricultural enterprises; the IACS register thus also contains information on industry code.

If some of the variables in a base register, which are often used to select or match, are out of date or incomplete, *register update questionnaires* should be carried out. Such questionnaires, which should be brief so as not to result in an unnecessary burden on the respondent, can be sent to the objects in the base register where data is missing or out of date.

Example: To maintain quality in the Business Register, two questionnaires are sent each year to all enterprises with more than one local unit. This questionnaire is actually carried out to maintain the quality of the population of local units, but the enterprises are asked at the same time about current industry and address details.

It is natural within the Business Register to spend more time maintaining register data on large enterprises that are of considerable economic significance in the enterprise surveys carried out by Statistics Sweden, but in a dynamic economy that is constantly changing; this approach can give rise to errors. The following example is given in Johansson (2001), where growth in the IT industry is studied.

Example: The little ones can be important!
Johansson describes growth in the IT industry between 1993 and 1998. The table below is based on data from the Business Register (Johansson, p. 82–83).

Chart 5.5 Employment within the IT industry by enterprise size category

Size Number of employees	0	1	2–4	5–9	10–19	20–49	50–99	100–199	200–499	500+	Total
Number of employees and self-employed, 1993	9 157	2 219	6 724	7 548	8 082	10 678	10 123	6 544	16 597	82 588	160 230
Number of employees and self-employed, 1998	17 825	3 084	8 459	9 634	11 582	16 434	13 206	12 933	21 045	85 600	199 802
Change 1993–98, number	8 668	865	1 735	2 086	3 500	5 756	3 083	6 389	4 478	3 012	39 572
Change in 1993–98, %	95%	39%	26%	28%	43%	54%	30%	98%	27%	4%	25%

Conclusion: Chart 5.5 above illustrates that the smallest enterprises have the largest growth in terms of number. This shows that it can be risky to disregard the small enterprises when maintaining a register. Important growth measurements could then be of low quality. The example above contains several important methodology problems – how can comparisons be made between years, how can employment and growth be measured? Comparability is a big problem for Johansson's survey, as the Business Register has level shifts in time series for 1996/1997 that affects the study of small enterprises.

Example: Population in the Business Register

Number of enterprises, according to Business Register		
	Old series	New series
1995	562 765	
1996	585 571	
1997	*601 385*	791 385
1998		810 337
1999		797 338

The register population in the Business Register consists of enterprises that are obliged to pay VAT, are registered as employers or that pay enterprise tax. Since 1996, an enterprise is obliged to pay VAT regardless of the size of its turnover; previously there was a cut-off limit of an annual turnover of SEK 200 000.

This new principle means that the number of enterprises increased by roughly 190 000 enterprises during 1997, to a total of 791 385, when the new enterprises that were obliged to pay VAT entered the register. The administrative VAT rules have therefore determined the definition of the register population. According to the example in Chart 5.4 above, the old series for agricultural enterprises had undercoverage of roughly 25%.

By combining several sources, the undercoverage could be reduced and the same statistically relevant population definition could perhaps be used both before and after the change in VAT obligations.

The coverage can vary during the year

A base register is based on sources that can refer to completely different periods and different time delays. The register's coverage will then vary during the year.

Example: Varying coverage in the Business Register

The Business Register is based on current monthly data on employers' VAT tax declarations, employer's charges and tax deductions for employees. However, for small enterprises that do not submit monthly tax declarations, the annual tax declarations that are submitted the year after the income year, are used.

The Business Register receives continuous information on new and restructured enterprises with varying time delays. Such new and restructured enterprises are particularly common at the turn of the year. The introduction of these changes takes both time and effort.

All base registers must fulfil the requirements we discuss in this section, even if we mainly use the Business Register as an example

5.4.7   Everyone should support the base registers!

The whole system should cooperate regarding the definitions of object sets and objects. Sample surveys and censuses take their frame populations, and the register-based surveys take their register populations, from the base registers. All those carrying out surveys on these populations gain knowledge, which should be forwarded to the team responsible for the base register so that the base registers can be the best sources of object sets within the statistical office.

The list below shows all the different information flows that contribute to good quality in a base register regarding objects and object sets. Those working with base registers should try to gain as much information as possible but, additionally, those working with other registers in the system should contribute with information that can improve the base register's quality regarding object sets.

1. Within every base register, contact should be kept with the *authorities* that deliver the administrative data. Changes in the administrative system can affect the register's objects and coverage.

2. Within every base register, it is possible to get information from *adjacent base registers*, where related objects can give information on the first register's objects.
   Examples: Changes in properties with residential housing must also signify changes of corresponding variables in the Population Register. If certain activities are missing for one person in the Activity Register, this can mean that the person is overcoverage in the Population Register.

3. Results from checks of different *statistical registers* can give information on coverage errors or incorrect object definitions.

4. *Sample surveys and censuses* can, via returned mail and reasons for 'no contact', give information on overcoverage and if the object has changed in some way.

5. *Register maintenance questionnaires* are carried out by a base register on parts of the object set where it is suspected that the data is incomplete or out of date.

The base register therefore plays a key role with regard to object sets and also regarding object definitions, which are discussed in the next section. So that the base registers can have the highest level of quality possible, all those using the base registers should forward information on any quality flaws discovered. There are three categories of users who should forward such information, discussed below.

Information from sample surveys and censuses

Those carrying out sample surveys or censuses have direct contact with data providers. Returned mail and tracking for telephone interviews can give information on overcoverage, out-of-date addresses, if an object has been restructured or other changes that are interesting for those working with register maintenance of the relevant base register. Household definitions in registers can be compared with sample surveys directed to households. Enterprise units in registers can be checked with different enterprise sample surveys.

Example: Overcoverage in the Population Register
Nonresponse in the Labour Force Surveys was analysed and gave an estimate of the number of foreign-born persons that were possible overcoverage in the Population Register. Returned mail and nonresponse in a questionnaire to foreign-born persons was also analysed and gave estimates of the overcoverage for various countries of birth, ages and sexes.

Information from primary registers

Those working with the primary registers receive the administrative registers. These can contain objects that are not in the base register. Furthermore, objects can have disappeared from the administrative register, which can indicate that the object has changed, been restructured or ceased to exist.

Example: Overcoverage in the Population Register
Those with no disposable income according to the Income Register were analysed further. The analysis gave an estimate of the overcoverage for different categories of foreign-born persons. It was shown that it is important to correct for overcoverage so that income statistics by country of birth are not misleading.

Information from integrated registers
Those working with integrated registers can compare variable values relating to the same object that come from different sources. If errors and inconsistencies are then detected, the reason for these could be that the object has changed.

Example: Incorrect object definitions in the Farm Register
The IACS register contains applications from agricultural enterprises for area subsidies. Wallgren (1999) matched this register against Statistics Sweden's Farm Register with the linkage variables PIN and organisation number. All hits were checked by comparing the area of arable land in the two registers. Roughly 9% of false hits were discovered. By combining units with the same telephone number, new units could be derived for which the area of arable land was correct. Object definitions are discussed in the next chapter.

5.4.8   Coverage problems in surveys with their own data collection

In surveys with their own data collection, the frame population is defined *before* the data is collected. Quarterly and monthly enterprise surveys have usually been based on the current stock version of the Business Register available in November. Surveys should refer to one reference month or one reference quarter during a particular year, which we will call year 2. The target population is therefore made up of enterprises that are active during the reference period in year 2. The frame population for these surveys is defined as the actual stock of active enterprises, according to the Business Register's version in November of year 1.

Chart 5.6 Overcoverage and undercoverage in surveys with their own data collection
Changes in the register population between November year 1 and December year 2
Per cent of frame population

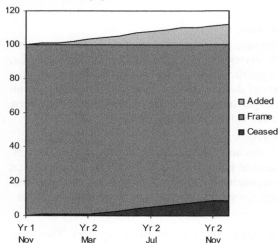

The enterprises that are added after November in year 1 give rise to *undercoverage*, which is marked in light grey in the chart.

The *remaining* part of the frame population is marked in medium grey in the chart.

The enterprises in the frame that ceased to exist after November in year 1 give rise to *overcoverage,* which is marked in dark grey in the chart.

Using the example in Chart 5.7 below, we show the problems with overcoverage and undercoverage that exist in surveys with their own data collection. The frame population that must be used is decided *before* the data are collected and this frame population is compared with the register population that is created *after* the survey period when all the administrative information has been received. The frame population that is defined in November in year 1 is used for annual surveys with their own data collection referring to year 1, as well as for surveys with their own data collection referring to quarter 1 of year 2.

The Business Register receives information on enterprise restructuring, newly started and closed-down enterprises after a fairly long delay. Furthermore, data on the enterprise's industrial classification can also be incorrect in the November frame, which will then be detected during year 2 or 3. At some point during the autumn after the year in question, nearly all the information regarding the previous year has been received and it is then possible to create a calendar year register of good quality.

If the frame population in Chart A is used for year 1, enterprise Idnr1 is overcoverage, Idnr 5 is undercoverage and the Industry for Idnr 3 is incorrect.

If the frame population in Chart A is used for year 2, Idnr 1 and 2 are overcoverage, Idnr 5 and 6 are undercoverage and the Industry for Idnr 3 is incorrect.

Chart 5.7 Frame populations and annual registers

A. Frame population formed in Nov year 1 for years 1 and year 2

Enterprise id	Industry
Idnr 1	DE
Idnr 2	DB
Idnr 3	**DA**
Idnr 4	DC
-	-
-	-

B. Calendar year register formed in autumn year 2 regarding year 1

Enterprise id	Industry
-	-
Idnr 2	DB
Idnr 3	**DB**
Idnr 4	DC
Idnr 5	DG
-	-

C. Calendar year register formed in autumn year 3 regarding year 2

Enterprise id	Industry
-	-
-	-
Idnr 3	**DB**
Idnr 4	DC
Idnr 5	DG
Idnr 6	DC

5.4.9 Coverage problems in register-based surveys

The main difference compared to surveys with their own data collection is that the register population in register-based surveys is created *after* the relevant reference period. Depending on how fast the administrative system receives the information that new objects have come into existence and that old objects have ceased, it is possible, after a time, to create a register population relating to a specific period or point in time. The coverage problems due to the frame population being defined *before* the reference period, which always occur in surveys with their own data collection, do not occur in register-based surveys. However, flaws in the administrative system can also result in certain categories of objects being missing or that important changes for certain objects are not reported.

Chart 5.8 Population definitions in different kinds of surveys

	Advantages	Disadvantages
Survey statistics, own data collection	Can be up-to-date	Significant problems with over- and under-coverage and errors in spanning variables if changes are reported late
Register-based statistics	Good coverage, more correct spanning variables	In certain cases, a long delay between the event to the statistics becoming available

A register population, created in the correct manner, has always better quality than the corresponding frame population, as it is based on more and better information.

5.5   DEFINING AND DERIVING OBJECTS

Statistical units or objects are synonymous terms, as a rule we are using the term object in this book. Object types in the register system can have been created in different ways:

1. In an administrative system for administrative purposes, such objects are often of legal importance.
2. In cooperation with an administrative authority for statistical purposes.
3. Within the statistical office by collecting information that has made it possible to define the objects.
4. Within the statistical office through the processing of the register, this category of objects is called *derived objects*.

5.5.1   Administrative object types

Many of the administrative object types are interesting for statistical purposes and the variables in these administrative registers can then also be of statistical interest.

Administrative object types that are statistically interesting:
Persons, jobs (employed or self-employed), properties or vehicles.

Administrative object types that are not always statistically interesting:
The concept of a family under taxation law (married/cohabiting with children in common) does not correspond to a statistically interesting household. The object type *legal unit* in an administrative enterprise register may not be statistically relevant, as one enterprise can report to the Tax Board using more than one legal unit identity.

5.5.2   Object types created in cooperation with administrative authorities

The statistical object types created by a statistical office in cooperation with an administrative authority have the advantage that certain administrative data can be linked to these objects of statistical interest; it is therefore possible to describe these object types with statistical variables without resource-intensive data collection.

Statistical object types created in cooperation with administrative authorities:
Local units are created by Statistics Sweden with information that has been collected by the National Tax Board. This information is collected via yearly statements of earnings from all employers. This is an example of integrated data collection, i.e. the national statistical office in cooperation with an administrative authority.

5.5.3   Statistical object types created via own data collection

Using surveys with their own data collection, it is possible to collect data on object types of statistical interest. To be able to describe these object types with statistical variables, resource-intensive data collection is required.

Statistical object types created using own data collection:
Housekeeping units and dwelling households are two kinds of object types created in household surveys. Within economic statistics, enterprise units, kind of activity units and local kind of activity units are formed after contacting the enterprises.

All administrative enterprise registers relate to objects called legal units, which are identified by organisation numbers. For statistical purposes, the object types enterprise units, local units, kind of activity units and local kind of activity units are formed within the Business Register. All these units have their own identity numbers and they must be observable in order for data to be collected. Extensive work is required to maintain a high level of quality.

5.5.4   Statistical object types derived by the statistical office

Statistical object types created within the statistical office by derivation are based on administrative objects. Administrative data for the administrative objects can then be used to describe the derived statistical objects.

Derived households

Households based on population registration of persons by dwelling as presently being done in Demark and Finland. If there is administrative data where all persons in the population are registered at a specific dwelling it is possible to define derived households as those who are registered at the same dwelling. This possibility is of central importance for register-based 'censuses'. Statistics Denmark (1995) and Statistics Finland (2004) describe how the traditional census information can be created by register-based surveys if it is possible to create derived households.

Derived enterprise units

All administrative sources give data concerning legal units. One enterprise unit can consist of one or many such legal units. To be able to use administrative sources for economic statistics, it is very important that derived enterprise units are created. Administrative sources give information about ownership and relations between legal units, and that information should be used to derive enterprise units.

In the chart below an example is given which shows how administrative data for one enterprise unit is reported to the Tax Board using different legal units. To be able to compare turnover and wage sum it is necessary to add values from all legal units belonging to this enterprise unit.

Chart 5.9 One enterprise unit which consists of seven legal units

		Turnover, SEK billions		Wage sum, SEK billions	
		Source 1	Source 2	Source 1	Source 3
Enterprise Unit 1	Legal Unit 1				0.1
Enterprise Unit 1	Legal Unit 2	8.6		1.3	
Enterprise Unit 1	Legal Unit 3		0.2		
Enterprise Unit 1	Legal Unit 4		1.8		0.2
Enterprise Unit 1	Legal Unit 5		4.2		0.6
Enterprise Unit 1	Legal Unit 6		1.7		0.3
Enterprise Unit 1	Legal Unit 7		0.9		0.1
	Sum:	**8.6**	**8.8**	**1.3**	**1.3**

5.5.5   Objects and identities – requirements for a base register

Besides the important role that a base register plays within the system regarding the object set, according to the Section 5.4.6, derived objects are also created within the base registers. Another important and difficult task is to follow objects and their identities over time, and to record statistically relevant changes.

Every object type should have an unambiguous and precise definition. When using administrative object types for statistical purposes, it is necessary to become familiar with the administrative system's definitions. Those creating objects within a statistical office are responsible for defining these object types. Both the administrative and statistical definitions should be documented.

When registers are matched, it is important that the *same identity* in different registers really *relate to the same object*. If matching produces false hits and variable values are imported from one object to another object, which has a similar identity, the survey results will be incorrect. The risk for such false hits is particularly great when matching registers from different points in time or periods, in which identities can have changed over time.

This type of problem exists in all parts of the register system, but to a varying extent. The same personal identification number in registers from different years can, in rare cases, refer to different persons. Changes in a personal identification number can also occur so that the same person has a different number in different annual versions of a register. An important task for those who are responsible for a base register is to maintain a cross-reference register with old and new identity numbers. The Swedish Population Register contains such cross-reference data on these personal identification numbers; when matching registers on individuals, this part of the Population Register should be used.

The same kinds of problems appear in the Business Register, where the same enterprise can operate under different identity numbers at the same time and these numbers can also be changed over time. A cross-reference register with different kinds of identity numbers should be maintained, and it is important to follow changes over time.

Real estate identities can consist of county, municipality, parish and such property codes will change when regional divisions change. This means that when matching registers from different years, keys should be used between the old and new regional codes.

The Activity Register is affected by both changes in personal identification numbers and enterprise identities. When the Activity Register and the Business Register are matched, it must be possible using local unit identities in the two registers to link together identical local units.

5.5.6   Objects and identities – requirements for primary and integrated registers

Those receiving administrative registers to create primary registers carry out the following processing so that the *objects* in the statistical register will follow the definitions specified:

– Identity numbers in the administrative registers are given the correct format. Duplicates in the administrative registers should be edited.

– The administrative register is matched against the base register containing the same object type. This match is carried out against the version of the base register referring to the same period. The identities of objects giving a mismatch are checked and corrected using, for example, name, address and telephone number.

– If the statistical register should relate to a different object type than that in the administrative source, it can be appropriate to either join together or divide up the values for the administrative objects. This is done using the base register, where the administrative objects are linked to the object type in question.

Example: In Chart 5.10 we have data (wage sums) describing administrative units (legal units, LeU). To describe wage sums for the statistical units (enterprises, EU) wage sums for all legal units belonging to the same enterprise unit are added.

Chart 5.10 Data for legal units are aggregated to describe enterprise units

Base register		Administrative register		Statistical register	
EU-id	**LeU-id**	**LeU-id**	**Wage sums**	**EU-id**	**Wage sums**
EU1	LeU11	LeU11	180	EU1	180 + 450 = 630
EU1	LeU12	LeU12	450	EU2	270 + 500 = 770
EU2	LeU21	LeU21	270		
EU2	LeU22	LeU22	500		

Note: The various economic variables are more or less well adapted to aggregation. If, for example, new orders are aggregated, orders to a sub-contractor will be counted twice.

Example: Chart 5.11 shows how it is possible to use a model to divide up turnover for one legal unit, LeU, by local units, LU. Here, the number of employees and industry turnover per employee for industry-specific enterprises are used as model. The choice of model is subjective – wage sums could have been used instead of the number of employees.

Chart 5.11 Data for legal units are divided up between local units

Base register				Model		Adm. register		Statistical register			
LeU-id	LU-id	Industry	Nr of empl	Industry	Turn-over/empl	LeU-id	Turn-over	LU-id	Industry	Empl	Model-calculated turnover
LeU1	LU11	DG	120	DA	2.1	LeU 1	300	LU11	DG	120	$\frac{300\cdot120\cdot1.5}{(120\cdot1.5+60\cdot2.1)} = 176.5$
LeU1	LU12	DA	60	DB	1.2	LeU 2	250	LU12	DA	60	$\frac{300\cdot60\cdot2.1}{(120\cdot1.5+60\cdot2.1)} = 123.5$
LeU2	LU21	DD	50	DD	1.8			LU21	DD	50	$\frac{250\cdot50\cdot1.8}{(50\cdot1.8+100\cdot1.2)} = 107.1$
LeU2	LU22	DB	100	DG	1.5			LU22	DB	100	$\frac{250\cdot100\cdot1.2}{(50\cdot1.8+100\cdot1.2)} = 142.9$

Those using data from several statistical registers in the system either to form an integrated register or to import data to their 'own' register should carry out checks to detect errors and inconsistencies regarding object sets and object definitions:

– It should be checked that the object sets in the different registers refer to the same point in time or period.

– Objects not receiving a hit when matched should be investigated – why was there not a hit? If all registers involved have been created with the respective base register, there should not be any mismatches, although there can still be missing values due to earlier mismatch with the base register.

– Check variable values from different registers, concerning related or similar variables, to find false hits.

– Inform those who are responsible of the registers concerned of any errors and inconsistencies discovered.

5.5.7   How to match – an overview

We have here been discussing *exact matching*, where the purpose is to find identical objects in different registers or to find objects that have a defined relation. There also exists *statistical matching,* where the purpose is to find *similar* objects for analytical purposes. When we are creating registers in a system of statistical registers, the matching method should be exact matching.

When creating a statistical register, data linking or (exact) matching is used for two different purposes:

– Different sources are combined to create an object set, the register population, with good coverage. This is discussed in Chapter 5.

– Different sources are used to create the variables in the new register. This is discussed in Chapter 6.

The matching process must be carefully planned – which linkage variables should be used, and in which order should the different sources be combined. The quality of the linkage variables is important, and editing of these variables is an important part of the work.

Causes and extent of mismatch should be investigated, and it must be decided if the non-matching units should be excluded or included in the register population. If they are included, mismatch will result in units with missing values for some variables. Seemingly matching objects should also be checked, since false hits will otherwise give rise to errors.

Example: Matching different farm registers.
Wallgren (1999) analysed matching problems connected with the IACS-register with applications for subsidies, the Farm Register and the Business Register.

The identifying variables should be edited before matching. Before editing of telephone numbers only 47% of the farmers in the Farm Register could be matched to corresponding units in the IACS register. After editing of telephone numbers 64% could be matched.

By combining two identifying variables (telephone number to the farm and the farms tax identity number) the matching result is improved, so that 96% of the units in the IACS register could be matched to units in the Farm Register. This means that by improving the matching methods it was possible to increase from 47% to 64%, and finally to 96% of successful matches.

Linkages must be checked. A match between identification variables is not sufficient proof that the IACS and Farm Register objects are identical. If the IACS object has a larger crop area than the Farm Register object, this can indicate that the IACS object should be linked with two Farm Register objects and vice versa. The linkages were checked by comparing total arable area, reliable crop area and location described by parish.

5.6   HOW TO PRODUCE REGIONAL REGISTER-BASED STATISTICS

Regional statistics at Statistics Sweden is created by a decentralized estimation process, which results in completely consistent and coherent micro data. The principles presented in this chapter are followed by those who work with a number of different registers with data concerning individuals. Due to the methods used it is possible to produce detailed and completely consistent regional statistics.

The first step in the process implies that a standardised population is created by the team responsible for the Population Register. This population is then the basis for those who work with the other registers.

The standardised population is defined as the population at December 31. The population for December 31, year t is created in early February year $t+1$. This standardised population is used as register population in the other statistical registers in Chart 5.12. The chart below illustrates the work:

Chart 5.12 Decentralized but coordinated processes to create registers on persons

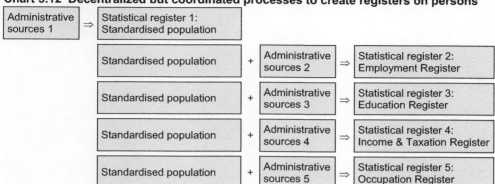

As the administrative sources 2–5 do not overlap regarding statistical variables, the work to create statistical register 2–5 can be done independently of each other. The five registers in Chart 5.12 can be integrated into one register with all variables.

Chart 5.13 Integrated register with parts from different registers on persons

PIN	Classification variables from the					Statistical variables from the …												
	Population Register					Employment Reg			Education Reg			Income & Taxation			Occupation Reg		
	var1	var2	var3	var..	var..	var..	var..	var..	var..	var..	var..	var..	var..	var..	var..	var..	var..
1																	
2																	
3																	
...																	
N																	

The regional classification variables in the Population Register have been created by adjoining regional codes and coordinates from the Real Estate Register to each individual in the Population Register. As persons are registered at the real estate where they live, this matching process is easy. The National Tax Board is responsible for this registration, and as it is important from an administrative point of view, where persons are registered, the quality of the regional codes and coordinates are good.

A large number of standardised tables are produced to meet the demands from different kinds of users. As coordinates can be used, the regional division is flexible. The risk for disclosure must be controlled, this is discussed in Chapter 12.

CHAPTER 6

How to create a Register
– The Variables

According to Section 5.1, the work to create a statistical register is divided into five different steps. During step 5c, the variable content of the register is created. What processing should be carried out to create the variables in question? The variables in the administrative sources should be checked and edited. Different sources are matched; variables are selected and imported to the new register and derived variables are created in the new register.

6.1   DECIDING THE REGISTER'S VARIABLE CONTENT

A variable is a measurable characteristic of an object. In Section 3.3, the concept *variable* is described both from the point of view of statistical science and informatics. In the register system, we distinguish between variables with different origins:

– *local primary variables* have been collected from an administrative register or from a census to the register in question;

– *locally derived variables* have been formed in the register in question using other variables in the register;

– *imported primary variables* are primary variables that have been brought to the register in question from another register in the register system;

– *imported derived variables* are derived variables that have been brought to the register in question from another register in the system.

6.1.1   Variable definitions

Every survey begins with a set of questions that is formulated in theoretical or general terms. The theoretical concepts in a set of questions must then be operationalised, i.e. translated into measurable concepts. When defining these measurable concepts, *what* is being surveyed is determined. We have previously mentioned this when discussing definitions of a survey's population. When talking about variable definitions, the theoretical concepts relating to the object's characteristics also need to be translated into clearly defined statistical variables.

A variable definition should specify which object type the variable relates to and the variable's reference point in time or reference period, i.e. that the characteristic relates to a specific point in time/period. In addition, the definition should specify how the characteris-

Register-based Statistics – Administrative Data for Statistical Purposes A. Wallgren and B. Wallgren
© 2007 John Wiley & Sons, Ltd

tic should be measured, and what measurement scale should be used. Two examples regarding age can be used to illustrate this:

Definition 1: Age refers to a person's age in full years reached at the end of the survey year. *Example:* A person born on 1 January 1980 has, for a survey referring to 31/12 in 2000, reached 20 years of age.

Definition 2: Age refers to a person's age rounded to one decimal place at the reference point in time. *Example:* A person born on 1 January 1980 has, for a survey referring to 31/12 in 2000, the age of 21.0 years.

The *definition of a primary variable* is determined when formulating a question on a questionnaire or administrative form. Instructions to the question are also included in the definition. When documenting statistics based on administrative data, questionnaires and brochures with instructions from the administrative authority should therefore always be collected and stored, preferably also electronically.

The *definition of a derived variable* is partly made up of the definitions of the variables that it is based on and partly of the rule for how the derived variable was formed.

For all kinds of variables it is also desirable that the definitions are both *adequate,* i.e. they agree with what is to be surveyed and *functional,* i.e. the definitions should be easy to use.

Adequate variable definitions

If a variable definition is adequate, this also means that *the definition agrees with other variables* in the register that are created for the purpose of the survey. In an enterprise register, definitions of variables such as production, sales, incoming and outgoing stock should match so that the logical relations between the variables are applicable. Production minus sales should be the same as the change in stock for the period.

Because all statistical registers are part of a system, *the variable definitions in the different registers should also match* for them to be adequate. For example, it is a basic rule in all registers on individuals that variables such as age, civil status, etc. are defined in the same way. Otherwise combined usage of register data would be made difficult and the quality of register-based statistics would be lowered due to a lack of consistency and comparability.

When qualitative variable values are grouped or quantitative variables are divided into class intervals, the groups or the class intervals should also be the same in different registers. For example, if age categories are the same in different registers on individuals, the published tables will be comparable, increasing the coherence of the statistics, which is an important part of the quality concept.

Functional variable definitions and derived variables

It is always functional to use the administrative variables' definitions. If the administrative variables are not sufficiently adequate for the statistical needs, an attempt should be made to form derived variables (see Section 3.3.3). To form variables of statistical interest, the possibilities presented by the register system should be exploited. By importing variables both from administrative and statistical source registers, a register with rich content can be created, and many interesting derived variables can be formed.

Example: The Income and Taxation Register (described in Section 1.4.1) is based on hundreds of tax administrative income variables. Using these, a number of important statistical variables, such as disposable income, can be formed.

Naming variables and standardised variables

Register variables that are used by several surveys or products should be *standardised.* This means that the responsibility for naming, for quality and for documentation of the variable lies with the register where the variable was first created at the statistical office. For this register, the standardised variable is therefore a local primary variable or locally derived variable.

When others use a standardised variable within the statistical office, the name and definition should remain unchanged. The documentation should be easily accessible for everyone. Those importing a standardised variable should not then need to produce their own variable documentation. The name of the standardised variable should not be used as the name for other variables.

6.1.2   Creating register variables

Once the new register's population has been created and edited, variable selection is carried out. Statistically interesting variables are imported from different sources.

1. The research objectives dictate what the content of the register should be. What are the users' needs? What possibilities for further projects are there?

2. Variables for the new register should be taken from all relevant sources. When creating a statistical register, any register data that is relevant to the research objectives should be used. The best possible variables should be created. The statistical register should be cleared of any obvious errors and should be consistent, i.e. not contain conflicting data. Other variables that are to be used for editing should also be imported. These variables for editing are correlated to the register variables so that edit rules can be formulated. Variables from previous versions of the register in question should be used for editing.

3. Adjoined derived variables are formed by matching against other registers. Aggregated derived variables are formed by matching and processing other registers. These types of derived variables are described in Section 3.3.3.

4. The register variables are edited. How can obvious and suspected errors be detected? Editing work is discussed in Section 6.3.

5. How should unreasonable and missing values be replaced? In Section 8.1 we describe how item nonresponse in registers can be dealt with.

6. Derived variables should be formed using the register's variables. This is dealt with in Section 6.2 below.

7. Register processing, the results of editing and the variables in the new register should be documented.

The registers created will be parts of the register system, which brings certain requirements relating to coordination and cooperation. The editing and processing should be adequate for other usage in the register system. This helps to avoid duplicate work and the value of the register that has been created is increased.

6.2   FORMING DERIVED VARIABLES USING MODELS

Derived variables are discussed in Section 3.3.3, where four types of derived variables are described:

1. Derivation of variables by grouping values or dividing into class intervals.

2. The derivation of a (statistical) variable using calculations and logical procedures with several (administrative) variables from the data matrix. This type of derived variable will be discussed in this section.

3. Derivation by adjoining a variable from another register referring to another object type. For example, industry sector for a gainfully employed person is established by giving the industry sector of the local unit where the person is employed. An adjoined variable can be multi-valued.

4. Derivation by aggregation of a variable in another register. For example, household income is established by summing up the values of the variable personal income for all the individuals in a certain household.

Using a number of administrative variables, it is possible to form a statistically meaningful variable. In this section, we use y for the derived statistical variable and x_1, x_2,... for the administrative variables. We distinguish between situations where y is a qualitative variable or a quantitative variable and between exactly calculated variable values and variable values estimated using a model.

Deriving variable values using calculations is related to *imputing* variable values. The difference is that the derived variable is created for *all* objects in a register, while the imputed variable value is only formed for the objects in the register where values are missing. Derivation and imputation are compared in Section 6.2.4. The discussions below are also of interest when calculating imputed values.

6.2.1   Exact calculation of values for a derived variable using a rule

In many situations, it is appropriate to form a new variable using a rule, which should be well founded and documented.

Example, qualitative variable:
The variable foreign-born or domestic born, y, is created in the Swedish Population Register using four other variables:

Chart 6.1 Classification of foreign-born and born in Sweden

y Foreign or domestic born Code:	x_1 person's country of birth	x_2 person's length of residence	x_3 father's country of birth	x_4 mother's country of birth
1.1 Foreign-born with residence 0–4 yrs	Foreign-born	0–4 yrs	-	-
1.2 Foreign-born with residence 5+ yrs	Foreign-born	5 yrs and more	-	-
2.1 Domestic-born with two foreign-born parents	Domestic-born	-	Foreign-born	Foreign-born
2.2 Domestic-born with one domestic-born and one foreign-born parent	Domestic-born	-	Foreign-born Domestic-born	Domestic-born Foreign-born
2.3 Domestic-born with two domestic-born parents	Domestic-born	-	Domestic-born	Domestic-born

Example, quantitative variable:
Disposable income y is calculated with a rule to show which income variables x_{i1}, x_{i2},... should be summed and which taxes x_{t1}, x_{t2},... should be subtracted.

In both the examples above, rules are formed for how the derived variable should be defined. These rules are based on subject matter knowledge and judgement. Calculations are precise, i.e. if the x variables do not have any measurement errors, the y variable should not have either.

6.2.2 Estimating the value for a derived variable with a rule

Rules for how the derived variable should be defined should also be used in the two examples below. The rules used are models that are completely based on knowledge of the subject field. However, the calculations here are not precise; for example, even if the x variables do not contain measurement errors, the y variable may contain errors. These errors in the y variable are called *model errors*. Section 6.2.3 discusses models that are based on causal analysis.

Example: Occupation in the public sector, giving priority to sources

The staff registers of public sector employers contain administrative variables used by Statistics Sweden to classify employees by occupation according to the standard for occupational classification (ISCO). The administrative variables *job title* and *TNS code* have been used according to certain rules. These rules were changed in 2000. The old and the new rules appear as follows when giving priority between the administrative variables in different ways:

Chart 6.2 Classification of occupations in the public sector

Until 2000	Since 2001
1st step: (ca. 90% cases) Applicable job titles exist and are used.	1st step: (ca. 48% cases) TNS code exists and is used.
Example: Job title: 93460 'web editor' becomes ISCO: 2451 'journalist, author, information officer, etc'	*Example:* TNS: 1316 'writing, testing and documenting programs' becomes ISCO: 3121 'data technician'
2nd step: (ca. 8% cases) No applicable job title but TNS code exists and is used.	2nd step: (ca. 51% of cases) NO TNS code but applicable job title exists and is used.
Example: TNS: 1316 'writing, testing and documenting programs' becomes ISCO: 3121 'data technician'	*Example:* Job title: 93460 'web editor' becomes ISCO: 2451 'journalist, author, information officer, etc'
3rd step: (ca. 2% cases) No useful job title or TNS code exist. Becomes nonresponse, ISCO missing	3rd step: (ca. 1% cases) No TNS code or applicable job title exist. Becomes nonresponse, ISCO missing

Even if the TNS code and job title are correct, the occupation according to ISCO can sometimes be incorrect. There is no exact relation between the two administrative variables and the actual occupation of an individual.

We can see here that the rules cannot be taken for granted; the administrative variables can be used according to different principles, and level shifts in time series can occur. Despite the ISCO classification remaining the same, level shifts in the time series can be caused when the rules are improved. If the quality of the TNS code is improved, it is appropriate that this variable is prioritised.

Correspondingly, there are rules to translate occupational codes from the municipalities, county councils and private employers to ISCO.

Example: Status of employment in the Employment Register 1985–1992.
In the first version of the Employment Register (1985), the intention was to use statement of earnings data to measure employment in the same way as with a traditional census – at least one hour of gainful employment during the measurement week in November. The statement of earnings data with information on all transactions between the employer and the employees was interpreted as follows:

– Statement of earnings for a part of the year including November: If total income per month (including sickness benefit) was higher than SEK 200, the individual was classified as gainfully employed in November.

– Statement of earnings for full year: if total annual income was higher than SEK 21 800, the individual was classified as gainfully employed in November.

– If none of the above conditions were fulfilled, the individual was classified as not gainfully employed in November.

– There were also special rules for sailors and self-employed persons.

Even if all variables from the statement of earnings data were correct, the above rule could still lead to incorrect classifications. Individuals who are gainfully employed according to the Population & Housing Census definition could be incorrectly classified as not gainfully employed, and vice versa. However, an evaluation was carried out and it was judged that the quality of the derived employment variable was acceptable. Between 1986 and 1992, these rules were used but with income limits adjusted by a salary index.

The statistical variable *status of employment* is based on three administrative variables: The point in time of the statement of earnings data (first and last month in the period the employment relates to), gross salary and sickness benefit. The rule states how these variables should be interpreted so that an individual can be classified as gainfully employed or not gainfully employed in November.

6.2.3   Estimating the value for a derived variable with a causal model

In both examples in Section 6.2.2, rules are formed for how the derived variable should be defined. These rules are based on knowledge of the subject field and individual judgement. An alternative to use such rules is to analyse the relation between the wanted variable y and the administrative variables (called here x_1, x_2...) by building a statistical (causal) model. Using that statistical model, the derived variable can then be created.

When creating derived variables using a statistical model, there are two steps involving different data matrices:

1. The first data matrix with *test data* from e.g. a sample survey containing both the y variable and the x variables. With this data matrix, a model is first put together to show how best to estimate y for the given values of the x variables.

2. The model is then used on the second data matrix, the *register's data matrix,* where only the x variables exist. With the estimated model, a y value for every object in the register is calculated, with the help of the object's known x values.

The advantage with a statistical model compared to a rule, based on knowledge of the subject field and judgements, is that a good statistical model shows how to best use many administrative variables. The model can contain many variables, as opposed to a rule based purely on knowledge of the subject field. The disadvantage is that it is necessary to generalise the analysis results from the test data in order to apply it to the register's data matrix –

that the model is good for the test data does not necessarily mean it is also good for the register's data matrix.

Quantitative derived variables

When y is a quantitative variable, there are three different types of models that can be used:

- *Group-related mean values*: The relation between the y variable and the x variables are studied in a table with the y mean values based on the test data.
- *Ratios:* If y can be assumed to be proportional to an x variable, the estimated ratio $\Sigma y / \Sigma x$ from the test data can be used. Different ratios can be calculated for different groups of objects.
- *Regression model:* If it is possible to build a regression model $\hat{y} = f(x_1, x_2, ...)$ with the test data, the estimated function $f(x_1, x_2, ...)$ can be used.

Example: Energy use in industry – group-related ratios

A sample survey is carried out among local units with 10–49 employees. It is then possible to have information on number of employees, industry and energy consumption for the sample that consists of around 1 800 local units. Among local units with *less than 10 employees* (ca. 50 000 local units), no sample is selected but it is assumed that, within each industry, the same amount of energy is consumed per employee as in enterprises with 10–49 employees.

Here, a sample of 1 800 enterprises makes up the test data that is used to calculate the ratios $\Sigma y / \Sigma x$ = energy consumption/number of employees for different branches of industry. These ratios are used to calculate estimated energy consumption (=estimated ratio · number of employees at local unit) for every local unit with less than 10 employees. A disadvantage of this is that a model that works for enterprises with 10–49 employees is not necessarily suitable for very small enterprises.

Qualitative derived variables

The example of the Employment Register above raises the question of whether the rule really uses the administrative variables in the best way. If, for example, the time information on the statement of earnings data is of bad quality, should there be a rule that is strongly dependent on this variable? By analysing test data from a sample survey, it is possible to study the relation between gainful employment and the statement of earnings variables.

If y is a qualitative variable, the following types of model can be used:

- *Regression model:* if y only has two categories, the relation can be studied using a regression model $\hat{y} = f(x_1, x_2, ...)$.
- *Discriminant analysis:* if y has more than two categories, a discriminant analysis model can be used to study how the x variables can best be used to classify the objects in the different y categories.
- *Data mining model:* software for data mining can also be used.

Example: Status of employment in the Employment Register from 1993.

The first version of the Employment Register used the following rules:

- Statement of earnings for the part of the year including November: if total income per month (including sickness benefit) was higher than SEK 200, the individual was classified as gainfully employed in November.

– Statement of earnings for full year: if total annual income was higher than SEK 21 800, the individual was classified as gainfully employed in November.

However, this rule-based methodology has serious disadvantages. Young and old persons are classified according to the same income limits despite the fact that their salary levels are different. Income patterns, the distribution between permanent and temporary jobs and the structure of the statement of earnings data have changed over time. The system for sickness benefits has also changed. All these changes result in that statistics from different years are not comparable, despite the same rules being used.

To solve the problems of comparability outlined above, a derived variable was introduced in the 1993 version of the Employment Register. For those persons who participated in the Labour Force Surveys (LFS) in November 1993, statement of earnings data were combined with status of employment according to the LFS. Using regression analysis, models were built for different combinations of sex and age for this test data. In this way, the different groups were given different income limits but all the limits corresponded to the employment definitions in the LFS. Another advantage is that administrative variables of high quality can greatly affect the classifications, while variables of low quality have little effect. A broad outline of the analysis is as follows:

1. With test data with known employment status according to LFS, regression models are estimated, where the employment status according to LFS is the y variable with two categories (gainfully employed/not gainfully employed) and the statement of earnings data are the regression model's x variables. The test data is divided into sub-groups, which have the same type of statement of earnings data, age categories and sex. Separate analyses are carried out for every sub-group.

2. Using the estimated model for a sub-group, an estimated y value is calculated using the x variables from the statement of earnings data. If the analysis succeeds, those classified as gainfully employed in the LFS will have estimated y-values that are markedly different from those not gainfully employed.

3. A cut-off value is determined so that those with estimated y values on one side of the cut-off value are classified as gainfully employed and the remaining persons are classified as not gainfully employed. The limit is set so that the number of persons classified as gainfully employed will be of the same size as the corresponding number according to the LFS in the test data.

4. These cut-off values for the different sub-groups are then used so that all persons in the register population will be classified using the administrative variables in the statement of earnings data.

For example, for the 2001 version of the register, the 2001 November LFS was used to produce new income limits via new regression analyses – these new limits have the same definition of gainfully employed in the LFS as previously. In this way, it is possible to carry out relevant comparisons between different years.

Qualitative derived variables that have been formed using a statistical model should be estimated so that *classification errors* can be judged. With a good statistical model, both *net errors* and *gross errors* should be minimal. Chart 6.3 below contains a comparison between the old and the new methods of defining employed persons in the Employment Register.

The estimated *gross error* in Chart 6.3 is an estimate of the share of incorrect classifications in the entire register, where the estimate of the *net error* is an estimate of the systematic error in the method of defining gainfully employed in the Employment Register (assuming that the LFS gives correct estimates). For good estimates of gross and net error, it is desirable that two sets of test data material should be used, one to build the statistical model and the other to estimate the classification errors.

Chart 6.3 Classification errors in the Employment Register 1993

Number of persons in test data	Estimate in **new** Emp. Register			Estimate in **old** Emp. Register		
	Employed	Not employed	Total	Employed	Not employed	Total
Employed LFS	22 360	1 158	23 518	22 472	1 046	23 518
Not employed LFS	1 068	6 872	7 940	1 329	6 611	7 940
Total	23 428	8 030	31 458	23 801	7 657	31 458

Per cent of total number of persons	Estimate in **new** Emp.Register			Estimate in **old** Emp. Register		
	Employed	Not employed	Total	Employed	Not employed	Total
Employed LFS	71.1	3.7	74.8	71.4	3.3	74.8
Not employed LFS	3.4	21.8	25.2	4.2	21.0	25.2
Total	74.5	25.5	100.0	75.7	24.3	100.0

Classification error						
	Net error:	74.5 – 74.8 = –0.3%		Net error:	75.7 – 74.8 = 0.9%	
	Gross error:	3.7 + 3.4 = 7.1%		Gross error:	3.3 + 4.2 = 7.5%	

6.2.4 Derived variables and imputed variable values

As we have previously mentioned, work with deriving variable values using calculations is related to *imputing* variable values. The difference is that a derived variable is created using calculations for *all* objects in a register, while imputed variable values are only calculated for those objects in the register that do not have a value. In the chart below, these two types of processing are compared.

Chart 6.4 shows a longitudinal enterprise register with wage sums for year 1 and year 2 (*Wsum1* and *Wsum2*), and the number of employees in year 1 and year 2 (*Emp1* and *Emp2*). Some values for the number of employees are missing. The imputed values are calculated by dividing the wage sum with the average wage sum per employee[1]. Then, two growth measurements are calculated as derived variables. *Wdiff* = 1 if wage sums have increased and *Ediff* = 1 is the number of employees has increased between years 1 and 2.

Chart 6.4 Imputed values and derived variables in a enterprise register

Before processing

ID	Wsum1	Wsum2	Emp1	Emp2
1	12 132	12 344	34	32
2	1 775	1 438	5	4
3	893	914	2	missing
4	18 923	17 835	53	47
5	239	346	missing	missing
6	6 221	7 583	17	20
7	549	514	3	2

With imputed values and derived variables

ID	Wsum1	Wsum2	Emp1	Emp2	Wdiff	Ediff
1	12 132	12 344	34	32	1	0
2	1 775	1 438	5	4	0	0
3	893	914	2	2	1	0
4	18 923	17 835	53	47	0	0
5	239	346	1	1	1	0
6	6 221	7 583	17	20	1	1
7	549	514	3	2	0	0

[1] The imputed value for enterprise 5 for the number of employees in year 1 is calculated as:
For the six enterprises for which values are known: Wage sum = 40 493; Nr of emp = 114
Imputed values rounded = 239/(40 493/114) = 1

6.2.5   Creating variables by coding

In some cases, data in text form is used to create statistically useful variables. The information is transformed from unstructured text to completely structured variable values in a coding process.

The location address of a local unit is the link between the Real Estate and Business Registers. There are several problems with these addresses that are currently not given in a standardised format. For example, the same street address can be given in different ways:

> Storgatan 17
> Storg 17
> Storg. 17
> and spelling mistakes can also occur: Storgtan 17

By processing addresses in a translation program, in which the actual addresses are compared with alias lists, many of these addresses can be transformed into a structured format (i.e. all variations become Storgatan 17). The addresses that cannot be clarified with such computer processing can then be coded manually. When both the property addresses in the Real Estate Register and the location addresses in the Business Register have the same structured format, the registers can be matched using the address variable.

There are several examples of important variables in the register system being created via coding. For some enterprises, industrial classification is determined by using the telephone catalogue's yellow pages. Causes of death and types of occupational injuries are coded using text information from administrative forms.

Causes of death
After investigation into the cause of death, a doctor produces a certificate on the cause of death. These certificates are scanned and registered in the computer. The diagnoses on these computer-registered forms are coded using special software that has been developed at Statistics Sweden. The software comprises the automatic coding of diagnoses from plain text to a code according to international statistical cause of death classifications. Uncertain cases are sorted and coded manually after contact with the relevant doctor.

Occupational injuries
When occupational injuries are reported, each employer is responsible for that a special form with information is sent to the social insurance office. The free text description of the occupational injury is coded by the statistical authority, the Swedish Work Environment Authority, into a number of statistical variables such as *incident, main external factor, suspected cause* and *diagnosis.* Coding instructions, training of the coders and coding checks are important phases in the process to ensure quality of the coding result.

6.3   EDITING AND CORRECTING REGISTER VARIABLES

In Chapters 5 and 6 we describe how administrative data are transformed into statistical registers. All the steps in this process contain editing of administrative data, but as we want the term editing to have a more precise meaning, we have chosen to use the term only for the editing work described in this section.

Own data collection and register-based surveys
The main editing phase for surveys with their own data collection involves editing of the collected data. It is often possible to contact the data providers to correct unreasonable

variable values. If the editing requires a large amount of resources, it can be a sign that the questionnaire needs to be redesigned. It is to be noted, that errors or suspected errors are interpreted as errors concerning variable values. The aim is to replace wrong or unreasonable values with corrected or reasonable imputed values. The scientific literature discusses as a rule only editing of data from surveys with their own data collection.

With register-based surveys, the data have first been edited by the administrative authority. After that, every administrative source is edited when the data has been delivered to the statistical office. But there is also another, more refined phase in the editing process where data from many sources (register-based surveys or censuses) are edited together. By this *consistency editing* it is possible to find further errors and inconsistencies. This consistency editing is an important phase that is missing in sample surveys with their own data collection. These aspects are illustrated in the chart below.

Chart 6.5 Editing in surveys with their own data collection and register-based surveys

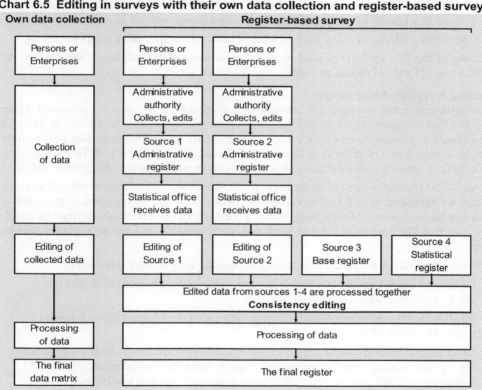

Errors in variables and errors in objects

In traditional editing of data in a survey where the statistical office has collected the data, all errors and suspected errors are interpreted as errors in variables. In consistency editing of data in a register-based survey we are editing data from different sources, and suspected errors can both be caused by errors in variables and errors in objects.

Errors in objects means that we believe that we compare data concerning the *same* object from different sources, however, the data we compare concern *different* objects, which erroneously have the same identity. This will be the case when we get false hits after matching, or when we have not created derived objects in a correct way. When we may have errors in objects we should not correct or impute variables values until we have checked that the objects are the same.

Example: Incorrect object definitions in the Farm Register (also in Section 5.4.7)
The IACS register contains applications from agricultural enterprises for area subsidies. Wallgren (1999) matched this register against Statistics Sweden's Farm Register with the linkage variables PIN and organisation number. All hits were checked by comparing the same variable; *area of arable land* in the two registers. In roughly 9% of the cases errors were detected, the area of arable land differed. These errors were caused by errors in objects. By combining objects with the same telephone number, new objects could be derived for which the area of arable land was the same in the two sources.

Our conclusion is that we must distinguish between different kinds of editing:

1. Ordinary editing of one source where errors are interpreted as errors in variables.

2. Consistency editing of many sources where errors are interpreted as errors in variables.

3. Consistency editing of many sources where errors are interpreted as errors in objects.

Editing of the first kind (1) is used in all kinds of surveys, but editing of the second and third kinds (2) and (3) should be used in register-based surveys.

Editing in register-based surveys
A sample survey has one main use, and only a limited number of tables are produced due to the fact that the sampling error will not permit detailed tabulation. The editing of data can be reduced to prevent 'overediting'; errors that don't affect the final estimates significantly can be overlooked. Overediting is discussed by Granquist and Kovar (1997). An overview of editing in surveys with own data collection can be found in Biemer and Lyberg (2003).

A statistical register is used in many register-based surveys and a large number of detailed tables are produced. It is therefore difficult to define what can be regarded as small errors that can be overlooked. Microediting is thus necessary, but editing methods must be developed so that the time used for this work will be reasonable and the data quality will be improved.

The role of editing is to increase the quality in both the long and the short-term:

– Detect and correct errors. Obvious errors can occur in the administrative data that can be corrected automatically.

– Identify the sources of errors and, in cooperation with the administrative authority, reduce the scope of these.

– Contribute to the increase in subject matter related expertise for the staff working with the register. Editing work helps the staff to get to know the characteristics of the administrative data and gain an understanding into how the data can be used.

To avoid duplicate work within the register system, the *local primary variables* should be edited at the register where the variable is created for the first time within the register system. Then, when these variables are exported to other registers, no new editing of each source should be needed, and consistency editing can be done directly.

The values that are missing, or that must be rejected, result in item nonresponse. These missing values can be replaced by imputed values. In Chapter 8 nonresponse adjustment in registers are discussed.

We present below three case studies that illustrate editing work of different registers at Statistics Sweden. These are followed by a summary of the results of the case studies.

6.3.1 Editing work within the Income and Taxation Register

As the first example for how editing work can be organised, we will use the Income and Taxation Register (I&T). This register is used to illustrate the distribution of income and taxation for individuals and families, using tax declarations and other administrative registers.

The register's variables are also used in the micro simulation model FASIT. This simulation model is used by the Ministry of Finance, among others, to study the effects of planned changes in taxation and transfer payments. The register has to fulfil many important quality requirements on a micro level, such as the income and taxation values for individual persons and families, which must be complete and consistent. To ensure that there are no strange simulation results, it is therefore necessary to carry out extensive work with editing and correcting the variable values.

A description of the work to create the register is given in Section 1.4.1, where the I&T register is used to illustrate how administrative registers are transformed into statistical registers. The different source registers used are given in Chart 1.4.

Step 1 – Register by register editing of administrative registers

In total, around 30 administrative registers are received every year and the first step is that every one of these is edited as follows.

Firstly, the *record descriptions* that come with the new administrative registers are checked. These record descriptions are generally changed every year, the variable name can have changed and new variables can have been added. The record description can be misunderstood, which will generate errors. It is therefore important to be in close contact with the persons at the administrative authority who can give the necessary explanations.

After this, the *extreme values* of the quantitative variables are studied. These are compared with the *previous year's values*, both on an aggregated and an individual level. Some variables, such as sickness benefit, have a ceiling value that can be used for these checks. Following this, *logical checks* are carried out to check whether the totals given are actually the sums of their parts. In all cases where aggregated values are an unexpected amount, where individual records have extreme values or where the totals do not match their parts, the administrative authority is contacted and the values are corrected. Errors can be caused by misunderstandings of record descriptions or that extreme values are due to a taxation measure that should not affect the statistics. For records on individuals, the error frequency is so low that it is possible to carry out record-by-record corrections in consultation with the administrative authority.

Certain variables are reported from subordinate authorities to the central authority, which in turn delivers the data to Statistics Sweden. In such situations, it is appropriate to check that all the subordinate authorities have provided data. For example, data on social assistance is usually missing for a few municipalities every year. It must be documented in the I&T register which municipalities have not provided data, and imputations can be carried out for these missing values, usually using the previous year's social assistance.

Editing work not only has a direct effect on the quality of the register. If the work is organised so that several persons share the responsibility, the editing can contribute to cooperation and the exchange of experience within the team. Through this, *subject matter expertise* is increased and, indirectly, even the quality of the register. When then documenting the work and the measures carried out to correct data, subject matter expertise is further strengthened.

Contacts with suppliers have several important effects. Firstly, it should be explained to the staff at the administrative authority how and for what purpose their data are used at Statistics Sweden. The staff at the authority should have an understanding for the consequences of lack of quality for the users of the statistics. For the subject matter expertise of the staff at Statistics Sweden, contacts with the suppliers are also important. This is why the staff working with the I&T register have regular meetings with the National Tax Board twice a year. These contacts are also used for the important work to identify new administrative sources.

The registers within Statistics Sweden's register system, which are used as sources for the I&T register should, in principle, not need to be checked again by the staff at the I&T-register – the checks should have been carried out on the primary register.

Something that has so far not worked particularly well is the exchange of experience between different units within Statistics Sweden. How many know exactly what is collected for the I&T register? Those who have data, that could be used for the I&T register's needs, do they know the requirements for the I&T? Does the Real Estate Taxation Register, for example, fulfil the requirements placed by the tax and wealth calculations in the I&T register? Lack of cooperation should not give rise to several units within Statistics Sweden receiving and checking the same administrative data. Duplicate work for Statistics Sweden and the authorities providing the registers should be avoided.

Step 2 – Final checking of the entire register
In the first step above, all the data from each authority are checked. In the next step, all variables from all sources are combined in one total register so that the different sources can be compared. All derived variables are then formed. In this way, new consistency checks can be carried out, i.e. that the sum of all variable values from different sources agrees with the sum from another source. In this way, additional errors can be identified.

This stage in the editing work is carried out on *a sample of records* in the total register. This is partly because the work otherwise would be too cumbersome and extensive, and partly because there are particularly strict quality requirements for the part of the total register that is to be used in the FASIT model. The total register consists of around 9 million records with 500 variables, which is why the sample method is a prerequisite to being able to carry out accurate checking. It is important to point out that the editing work increases the subject matter expertise of the staff – working with data at micro level can provide important skills – working with a sample can be a good way of getting to know the data.

Not so many errors are usually found in this way, so they can be corrected manually in consultation with the administrative authorities. If the corrections in the sample do not have so great an effect on the estimates in the sample, the total register can be approved and the checking phase brought to an end. If the corrections in the sample are more extensive, it could be necessary to use some automatic checks and correction of all records in the total register. However, such methods are currently not used.

Step 3 – Checking of estimates
In this step, all important tables are formed using the whole register as the basis. Estimates are checked and compared with the previous year's values. Furthermore, a number of simulations are carried out using the FASIT model, for the sole purpose of testing the data quality. If, for example, housing benefit remains unchanged in the model, then the model should generate model values that agree with the previously produced tables.

6.3.2   Editing work within the Statement of Earnings Register

The Statement of Earnings Register is used to calculate industry-specific wage sums, and is also used when the Activity Register and the Employment Register are created. This section gives an account of the editing work carried out on the definitive statement of earnings data for income year 2001, which was received by Statistics Sweden up to October 2002. The statement of earnings data are checked by those responsible for the Statement of Earnings Register and the edited register is then used as the source for other registers within the register system.

Checking of population definitions

The first step in the editing process involves checking that the number of received statements of earnings agrees with those sent from the National Tax Board.

The second step is to create a data matrix with the final statement of earnings data according to all the amendments in the consignment. The National Tax Board does not change input data – when the data provider (in this case the employer) submits amendments to them, new records are created equivalent to deletion, amendment or replacement of previous records. Because of this, processing is required in the register to take out invalid records and to check for duplicate records. The variable values for around 10 300 records are carried across from the original statement of earnings to the amendment record, as the amendment record can be incomplete.

The third step in the editing work is to check all identities. As statements of earnings can contain individual and enterprise identities, both personal identification numbers and organisation numbers should be checked. Around 7 600 personal identification numbers were incorrect, of which 5 000 could be corrected automatically.

The fourth step involves matching the personal identification numbers in the statements of earnings with those in the Population Register for 31/12/2001 and matching enterprise identities against the Business Register for March 2002. In both cases, several mismatched records are found – the Statement of Earnings Register contains more personal identification numbers than the Population Register and roughly 12 200 more enterprise identities than the Business Register. As the statement of earnings data relate to the full year, it should be matched against *calendar year versions* of the Population and Business Registers. The 12 200 records that were not in the Business Register consist mainly of identities that are used temporarily for bankrupt enterprises, etc. and, to a certain extent, of incorrect personal identification numbers.

Checking of variable values

In the *fifth step*, deviation errors are checked using 16 different probability checks. The relation between earned income and tax is used in several ways, plus that a search is made for records with extremely high earned income or tax. Around 5 000 records with extreme values are detected from these checks. These are checked in a simple way and only a few are checked with the National Tax Board. After these checking stages, each statement of earning is accepted, replaced by a new statement or taken out of the register.

Checking of the most important variable

The most important phase in the editing work involves checking that employed persons are linked to the correct local units. This link is crucial for the whole register system as it makes it possible to report gainfully employed persons by industry sector and region. Difficulties arise with this link when enterprises have more than one local unit. In spite of the fact that the employer has a duty to indicate the local unit on every statement of earnings, this information is often missing and sometimes unreasonable. Unreasonable local unit numbers are identified by comparing the number of employees with corresponding data in the Business Register and with data from the previous year's version of the Statement of Earnings Register. Plausibility in terms of commuting distance is also considered.

When a local unit is missing, or appears unreasonable, on the statement of earnings for enterprises with more than one local unit, the employer is contacted via a special data collection using a register update questionnaire. Those responsible for the Statement of Earnings Register work together with those responsible for the Business Register to catch changes regarding the local unit's municipality code and industrial classification code.

Output editing

The Statement of Earnings Register is used as a source for the Employment Register and, by checking the output from the Employment Register, the quality of the Statement of Earnings Register is also checked. Detailed tables with employed persons by industry sector and municipality are put together and compared with the previous year's tables. Deviations are checked and the results of these checks are documented. This documentation is very useful as many users inquire after publishing and question the results. Where documentation exists, those who are in contact with the users can respond that 'we have checked and the results are correct as far as we can see!'

6.3.3   Editing of income declarations from enterprises

The FRIDA register is based on income declarations from enterprises with taxation variables. This register is used in a similar way to the FASIT model in Section 6.3.1, but the aim is to study tax effects on enterprises and self-employed persons. The FRIDA register has been developed over several years in close cooperation with the Ministry of Finance. The register is documented, the editing methodology is described and the administrative questionnaire types are included in the documentation.

The original idea was to create one register with all enterprises. But it was discovered that the quality of the data was too poor for this to be possible. The editing work would have been too burdensome to create a total register of good quality. The quality of the data over the years has improved, but above all, the editing program that has been developed is so advanced that the creation of a total register could now be considered. One prerequisite for this register to be used for simulation models is that the editing should be so effective that the quality of the data is high. The high demand for quality is one of the reasons why it has been decided to work with definitive income declaration data and to compare these with data on the decisions taken by the Tax Board that becomes available at the same time. The complicated kind of data has meant that long-term efforts have been made to build up subject matter expertise and to maintain close contacts with both the administrative authority and the customer.

The first step in the work is the receipt of the data consignment from the administrative authority, in this case the National Tax Board. The consignment is extensive and the data does not have the structure that is usual within Statistics Sweden, having instead a variety of sub-records: *identity, record type, value.* Data are sent by variable and not by row or by object. The transfer of data can take over a week; the data consists of around 1 million questionnaires with between 300 and 400 variables.

The next step is to restructure all data into data matrices, with all the variables relating to the administrative questionnaire from one enterprise in one row in the data matrix. One data matrix is created for each type of administrative form or questionnaire. Thorough checks are necessary to check that all the data are included and interpreted correctly when restructured. Administrative forms and questionnaires change between the years, requiring contact with the National Tax Board and access to all questionnaire types for the year. These questionnaires are complicated with many economic and taxation concepts, which require a high level of subject matter expertise in the unit receiving income declarations from enterprises, so that the material is interpreted and used in the right way.

After having received and structured the data in the same way as described above, the next step is to create *derived enterprise units* of individual business owners. This enterprise group is difficult to work with, with several types of questionnaires, and many self-employed persons with several different activities. All enterprises that can be linked to persons in the same tax assessment household form one enterprise unit.

A first check is carried out on the total set of data to find specific technical errors. One such error is that the box number on the questionnaire has joined with the amount. If, for example, a reported result of SEK 1 000 000 was given in box '605', this amount could incorrectly have become 6051000000. The checking program should therefore be able to recognise the pattern '605'. These errors are then corrected manually.

The population of enterprises is stratified by enterprise type and size and then the sample is selected for the FRIDA register. Inactive enterprises are then deleted and the records that have not been defined as inactive are then run through the automatic checking and correction program.

Incorrect totals, typing errors and taxation changes that have not been corrected in the data from the National Tax Board are common errors in the material. The checking and correction program checks all totals and also that the values that are moved from one box in the questionnaire to another match. The different types of errors that those working with FRIDA have detected over the year have been added to the program and correction is carried out automatically. If a total matches after amendment according to a specific error type, then the correction is made according to this type. In this way, symbol errors, 1000 errors and other errors are remedied.

A sub-sample consists of roughly 40 000 limited companies. Of these, around 30 000 enterprises go through the checking program without any changes, as all the totals match. But errors are detected for around 10 000 enterprises. Of these, around 8 000 can be corrected automatically using all the error types in the program. The remaining 2 000 is further checked. Of these, around 1 700 have small errors, in terms of percentage, so they are approved without further action. There therefore remain around 300 enterprises with significant errors that are checked manually. These 300 are compared with the enterprise's annual report that has been submitted to the Patent and Registration Office. Records for around 60 enterprises can then be corrected but the remaining 240 cannot be remedied. For every incorrect total, a specific variable is created where the difference is stored. Every correction made is allocated a correction code. Records containing errors that cannot be corrected are allocated an error code.

Tax regulations change so that the data changes every year and variables disappear while new ones are added. This means that the editing program must be modified on a regular basis. Furthermore, the users' needs also change, which means that any changes in the editing must be discussed with the users so that the work at Statistics Sweden is focused on the most important problems. Because the staff at Statistics Sweden 'live with data at micro level', a learning process is always underway which leads to better subject matter expertise.

It is important that the staff at a statistical office also uses the data and regularly discuss users' problems.

6.3.4   Summary – editing processes in the register system

An analysis of the three cases shows that the editing of data from administrative sources can be structured in the following six stages. Under every title, we summarise the case descriptions given above:

1. Create a data matrix and *combine all records* that belong to the same object.

| **Statement of Earnings Register:** |
| A data matrix is created with the final statement of earnings data after all changes received in the consignment. Because of this, processing is required in the register to take out invalid records and to check for duplicate records. |

2. Editing should be carried out to *check the register population*, i.e. that no objects are missing or that there are no foreign objects or duplicates. This is done by checking against the relevant base register. This also includes the checking of all identity variables and that they are given the correct format.

| **Statement of Earnings Register:** |
| Around 7 600 personal identification numbers were incorrect, of which 5 000 could be corrected automatically. Personal identification numbers in the statement of earnings data were matched against the Population Register for 31/12/2001 and the organisation identity against the Business Register for March 2002. In both cases, there were many mismatched records. |

3. Editing can even be necessary to *check the object*, i.e. that the data regarding a specific identity from different sources refer to the same object. This can be done by comparing values for similar variables from different sources.

| **FRIDA, income declarations from enterprises:** |
| Derived enterprise units regarding individual business owners are created. All enterprises that can be linked to persons in the same tax household form one enterprise unit. |
| **Section 6.3, first example:** |
| The IACS register was matched against Statistics Sweden's Farm Register. All hits were checked by comparing the same variable; area of arable land in the two registers. In roughly 9% of the cases errors were detected, these errors were caused by errors in objects. New objects could be derived for which the area of arable land was the same in the two sources. |

4. It should be checked that the *delivery is complete*, both regarding objects and variables. Many administrative variables are only given for the objects concerned; if the value is zero, there is no administrative data. By this reason it is difficult to differentiate between missing values and true zero values. The incomplete delivery of variables is investigated by checking tables in which the totals or frequencies are compared against the previous year's values. Totals/frequencies are calculated for suitable groups, such as municipalities, so that incomplete deliveries can be detected.

| **Statement of Earnings Register:** |
| The first step in the editing process involves a check that the number of received statements of earnings agrees with those sent to the National Tax Board. |
| **I&T Register:** |
| For example, data on social assistance is usually missing for a few municipalities every year. It must be documented in the I&T Register which municipalities have not provided data, and imputations can be carried out for these missing values. The previous year's social assistance is usually used for this. |

5. Checks should be carried out to *check the variable values* so that obvious or suspected errors can be detected and corrected. Misinterpreted record descriptions and technical errors can result in *obvious errors* that can be easily corrected.

I&T Register: Record descriptions are generally changed every year, the variable names can have changed and new variables can have been added. The record description can be misunderstood, which will generate errors.

The extreme values of the quantitative variables are studied. These are compared to the previous year's values.

Logical checks are carried out that the totals really are the sum of their parts. Errors can be caused by misunderstandings of record descriptions or that extreme values are due to a taxation measure that should not affect the statistics.

Checking of prepared estimates: In this step, all important tables are formed using the whole register as the basis. Estimates are checked and compared with the previous year's values.

Statement of Earnings Register:
Deviation errors are checked using 16 different probability checks. The relation between earned income and tax is used in several ways, plus that records with extremely high earned income or tax are searched for. Around 5 000 records with extreme values were found after these checks.

After these checking stages, each statement of earning is accepted, replaced by a new statement or taken out of the register.

Detailed tables with employed persons by industry sector and municipality are put together and compared with the previous year's tables. Deviations are checked and the results of these checks are documented.

6. The editing process should be *documented,* by informing firstly on which methodology is used and secondly on the scope of different types of error and how these have been corrected. The results of checks should be documented, both using error and correction codes in the completed register and with a description of the effect and scope of the checks.

Statement of Earnings Register:
Detailed tables with employed persons by industry sector and municipality are put together and compared with the previous year's tables. Deviations are checked and the results of these checks are documented.

This documentation is very useful as many users inquire after and question the results. Where documentation exists, those who are in contact with the users can respond that 'we have checked and the results are correct as far as we can see!'

FRIDA, income declarations from enterprises:
For every incorrect total, a specific variable is created where the difference is stored. Every correction made is allocated a correction code. Records containing errors that cannot be corrected are allocated an error code. The register is documented, the editing methodology is described and the administrative questionnaire types are included in the documentation.

6.3.5 What more can be learned from these examples?

The above examples show that the administrative data received at Statistics Sweden can contain errors that require checking at micro level. Once these errors have been detected, they are often simply corrected. The demands to be fulfilled by the checking procedure depend on how the register is to be used. Statistics Sweden's statistical registers are often used for research, and for such advanced analytical needs, the quality on a micro level needs to be higher than when only simple tables are produced and higher demands are placed on the checks. High demands are primarily placed with regards to longitudinal studies when the linkages on micro level must be reasonable.

Subject matter expertise and contacts with suppliers
An overall conclusion is that subject matter expertise is of great importance for the effectiveness of the editing and checks. For surveys with their own data collection, it is sufficient

to be familiar with the survey in question, which is rarely changed. With register-based surveys, however, it is necessary to be familiar with the administrative system that has generated the data. Such a system can contain many complicated variables that are changed often.

Editing work not only has a direct effect on the quality of the register. If the work is organised so that several persons share the responsibility, the editing can contribute to cooperation and the exchange of experience within the team. Through this, subject matter expertise is increased and, indirectly, also the quality of the register. When then documenting the work and the measures carried out to correct data, subject matter expertise is further strengthened.

The method of working closely with a sample from a large register could be used more generally. In the case of the Income and Taxation Register, it is natural to work with the sample used in the micro-simulation model FASIT, but other registers could also use this method to test the quality and get to know the data material, by working with data analysis in a way that is not possible with the complete register.

The example also shows the significance of cooperation and development of expertise within the working group that receives the administrative registers, and the importance of having good contacts with the authorities supplying the data. Furthermore, co-operation between different teams working with related registers should be encouraged so that the administrative data is used effectively.

If the staff at the statistical office 'live with the data at micro level', the learning process is ongoing, which leads to better subject matter expertise. This learning process is strengthened by close contacts with users.

Duplicate work and exchange of experience
Duplicate editing work can occur; the editing of income declarations from enterprises, could be carried out in co-operation between several units at Statistics Sweden. Co-operation could also exist in relation to real estate assessment data. As the editing of register data has not yet attracted much attention, the important exchange of experience that is essential for the improvement of this aspect of register-statistical methodology work is not yet taking place.

Software for automatic editing and imputation
Computerised editing routines should be developed to detect and correct errors in our extensive registers. The development of methods for editing and correcting microdata in large registers is needed – the methods for the macro editing of sample surveys are not sufficient for the needs of register-based statistics. The example of FRIDA (income declarations from enterprises), can here be used for inspiration. The editing and correction program used by the FRIDA register is based on extensive subject matter expertise that has been developed successively.

Within the EU project EUREDIT, reported by Pannekoek and de Waal (2005), methods for completely automated editing and imputation have been developed. It remains to be examined whether these types of program software can be used to edit administrative data. However, automation should never obstruct the development of subject matter expertise.

According to de Waal and Quere (2003), only a small number of countries (not including Sweden) have developed advanced editing routines that use automatic editing and imputation based on the Fellegi-Holt (1976) paradigm. Such methods could be used to edit administrative sources one by one, but the consistency editing of many sources simultaneously, mentioned in the beginning of Section 6.3, gives rise to more methodological problems as errors can be caused both by errors in variables and errors in objects.

Example: Automatic editing and imputation of income declarations from enterprises.
This example illustrates automatic editing and imputation, which is used when the first editing of administrative data is done with one source at a time. In total 464 567 income declarations for 2004 were received from small enterprises using the special tax form for small enterprises.

Special software has been developed for this tax form, and the methods used are based on subject matter expertise. The last two variables in the table below, 'Adjustments' and 'Income for taxation' have the highest quality as they are of great legal and tax administrative importance. The software starts with theses values and then searches for errors in positive or negative signs and summation errors. In this way it is rather easy to find and correct most errors in this register with enterprise data.

Chart 6.6 shows that data is inconsistent before editing:

Receipts – Costs – Depreciations etc.
do not sum into total Income (40.734 –665.016 – 2.877 ≠ – 8.825)

The main reason for these errors is the cost C5; if seven records are corrected for errors in this variable, almost all inconsistencies in the register with 464 567 records disappear.

Also the costs C3 and C4 should be corrected; 16 corrections will change costs with almost 13 billions.

There are 12 855 small errors in the variable D6, if these are corrected or not is not important in this case, as the total sum of D6 is almost unchanged.

Chart 6.6 Automatic editing and imputation

		Before editing SEK billions	After editing SEK billions	Number of corrections
Receipts				
	R1	31.017	30.793	115
	R2	9.323	9.315	23
	R3	0.394	0.392	4
Receipts, total		**40.734**	**40.500**	
Costs				
	C1	-8.845	-8.833	22
	C2	-0.913	-0.913	0
	C3	-10.363	-0.961	3
	C4	-6.871	-3.110	13
	C5	-628.046	-3.027	7
	C6	-2.254	-2.252	6
	C7	-7.725	-7.602	23
Costs, total		**-665.016**	**-26.676**	
Depreciations etc.				
	D1	-4.078	-4.097	25
	D2	3.880	3.339	4
	D3	-3.217	-3.216	3
	D4	-0.975	-0.957	1
	D5	0.905	0.905	1
	D6	0.607	0.751	12 855
Depr. etc., total		**-2.877**	**-3.252**	
Income		**-8.825**	**10.572**	
Adjustments		**-17.789**	**-17.628**	
Income for taxation		**-7.053**	**-7.056**	

The conclusions from this example are:

- Administrative data must be edited, as there are many records, automatic software should be used.
- In many cases, a small number of huge errors destroy data. As a rule it is easy to find and correct these errors.
- Subject matter expertise should be used when the software is developed.

Base registers – suitable version does not exist

All of the three cases described above need *calendar year versions* of the Population Register and/or the Business Register to define their register populations. These register versions did not exist when the editing was done in these cases and this caused unnecessary mismatches. According to the principles described in Section 5.4, an attempt should be made to create the best possible calendar year version using all sources available once the income declarations from enterprises have been delivered. The calendar year version should contain all enterprises that have carried out activities at any time during the reference year.

Additional data collection can be necessary

When a variable in the administrative data is seen to have too low a level of quality to be used for statistical purposes, it can be necessary to conduct additional data collection to get a sufficiently high level of quality. The editing work with checking that employed persons are linked to the correct local units in the Statement of Earnings Register is one example of this. To achieve sufficient quality some employers are contacted via a special data collection using a register update questionnaire.

Editing as the basis for the quality declaration

If the register's staff are 'living with data at micro level', they obtain a good idea of the register's quality. This knowledge is a good basis for writing quality declarations.

6.4   CREATING LONGITUDINAL REGISTERS

Kardaun and Loeve (2005) describe and compare longitudinal analysis in some statistical offices. Register-based longitudinal surveys are common in Scandinavian countries, and surveys based on data collection are common in Anglo-Saxon countries. They also report that most longitudinal surveys are person oriented and only a few are business oriented.

In Section 1.6 we mention that register-based surveys have the advantage that they have complete coverage of time. Due to this fact, register-based surveys are suitable for longitudinal analysis. But this way of using administrative data requires that the longitudinal register has been created so that the longitudinal quality regarding objects and variables is sufficiently high.

In Chart 2.10 the system of statistical registers is illustrated. In the chart there are five longitudinal registers on persons and one register on enterprises or local units.

What is the statistical register to be used for? The use of the register determines the quality requirements that must be fulfilled, and this is the determining factor for which type of editing is needed and which type of register processing must be carried out.

A register that is only used to produce annual official statistics should provide estimates of good quality on an aggregated level. However, if the register is also to be used for longitudinal analyses, the *objects* or *statistical units* must be defined so that they can be followed over time without statistically unimportant changes disturbing the patterns.

Example: If an enterprise changes legal form from a trading company to a limited company, this should not be interpreted as the closure of one enterprise and the formation of new one. A new statistical product at Statistics Sweden, *Dynamics of enterprises and local units* aims to provide a more detailed chart of structural changes in the business sector. The aim with the register is to improve the classification of events within enterprises, presenting new possibilities to study the mobility of the labour force.

Primarily, stable identities for enterprises are created; allowing us to follow them over time in a more analytical way than can be done in the Business Register. With the Dynamics of enterprises and local units Register, newly constructed enterprises, closures, divisions and mergers are followed at the same time. As indicators of change, the staff of an enterprise is primarily taken into consideration at different points in time while less importance is given to change in owner, industry or location.

The basic rule in the Dynamics of enterprises and local units Register is that if the majority of those employed in year 1 also make up the majority of those employed in year 2, the units for the two years are considered as the same enterprise, regardless of the organisation number. Correspondingly, by setting up other rules regarding the flow of staff between the two years, it is possible to classify mergers and divisions of enterprises.

In the same way, the *variables* must also be defined so the changes over time can be statistically meaningful. When administrative rules change in the area of taxation, for example, the variables affected will not be comparable over time.

Example: The rules for taxation of reported profit from private limited companies has changed several times. The rules influence whether the company's owner takes the surplus as profit or as salary. Statistics containing wage sums and operational surpluses are therefore affected by this. The administrative variables *salary to enterprise owner* and *reported profit* should therefore be combined into one statistically meaningful variable for these private limited companies.

Even the variables formed within a statistical office can be more or less appropriate for longitudinal analyses. The method used to form the variable *gainfully employed persons in November* in the Employment Register, as described in Section 6.2.3 above, can lead to fictitious changes for persons close to the income limits that are used in the statistical model, which is the basis for the derived variable.

Chart 6.7 Fictitious changes in employment status for an individual

	Year 1	Year 2	Year 3	Year 4	Year 5
Income for individual every year	47 400	48 500	49 600	52 800	53 900
Limit in model every year	47 000	48 600	49 400	52 100	54 200
Status of employment, estimated value	Empl	Not empl	Empl	Empl	Not empl

With this type of derived variable in the Employment Register, estimates can be produced on an aggregated level to allow good quality comparisons over time. However, if the same variable is used for longitudinal analyses, where the change in employment status is studied for a group of individuals, fictitious changes can mean that the quality is not sufficiently high. The share of employed persons can be estimated with good quality, but it may not be possible to estimate the share of persons who have changed employment status with sufficiently high quality.

The advanced uses of statistical registers, such as longitudinal analyses, simulation and forecasting models place high demands on editing and processing. The rules for defining and checking objects must be adapted to suit the needs of these uses. Furthermore, the longitudinal register needs to be supplemented with a number of new imported variables and new derived variables that often require extensive processing.

The processing that is carried out in relation to advanced uses must be documented in detail. This also applies to register processing carried out for individual projects for researchers. This documentation is needed for methodology discussions and to give ideas for future tasks and other products.

CHAPTER 7

Estimation Methods

After having carried out the processing described in Chapters 5 and 6, the register's data matrix or data matrices are ready for use. The next step is to use the data matrix to form the statistical tables that are relevant to the research objectives in question. In this and the next two chapters, we describe the estimation methods that are used or that could be used to form estimates and tables.

We discuss some quality-related problems and give suggestions for solutions of these problems, based on certain estimation methods. Some of these estimation methods are based on the principle that weights are also used for register-based statistics, in a similar way as for sample surveys. Such weights are a special type of derived variables.

When using the data matrix to create statistical tables, the table cells will contain frequencies, sums or other statistical measures. When weights are used for estimation, weighted frequencies or weighted sums are calculated. In this chapter we give a general introduction to estimation methods used in register-based surveys.

In Chapter 8, we describe estimation methods that can be used to handle problems with *missing values, overcoverage* and *level shifts in time series*. These quality issues are as a rule not dealt with, when register-based statistics is produced, but the methods in Chapter 8 can be used to compensate for these sources of error so that the errors are reduced. The estimation methods discussed in Chapter 8 build on weights, calibration of weights and imputations.

In Chapter 9, we describe estimation methods that can be used to handle quality problems which arise when data from different registers are integrated. Such integration can give rise to *multi-valued variables* and special kinds of errors and inconsistencies. Multi-valued variables (mentioned in Section 3.3.2) are common in register systems where data from different kinds of objects are integrated.

As the registers in the register system interact, missing values and other quality problems in one register will also affect other registers that import data from the first register. Even the method of nonresponse adjustment chosen for one register affects the other registers in the system. The methods we propose must therefore function within the whole system so that the statistics from different registers are consistent.

The methods in Chapters 7–9 are illustrated using examples. These examples have been simplified so that the general principles are clear. Some examples are based on actual Statistics Sweden data, while other examples are based on small fictitious sets of data.

7.1 ESTIMATION IN SAMPLE SURVEYS AND REGISTER-BASED SURVEYS

The term *estimation* is generally used for sample surveys, but should also be used within register-based statistics. It is also important to distinguish here between the actual values in the target population and the estimates produced with the register.

7.1.1 Estimation methods that use weights

All formulas below refer to a particular cell in a table; the notation is simplified by this restriction. In a cell in a table (shaded in the table below) there are R observations from the register and we want to estimate the cell total Y in the register population. This cell total must in some cases be adjusted due to quality problems.

		$y_1, y_2, \ldots y_R$		

For every cell in a table, *sums* are calculated for quantitative variables, such as salary or turnover using formulas (2) or (3) below. If the *frequencies*, or number of observations, in the table cells are to be calculated, the variable $y = 1$ is used in the formulas for all observations.

With *survey samples*, estimates are made using formula (1) shown below. The design weights d_i depend on how the sample has been designed or allocated into different strata. The weights g_i in formula (1) are based on the auxiliary variables from statistical registers and are used to minimise sampling error and errors caused by nonresponse. Deville and Särndal (1992) introduced this method of estimation.

$$\hat{Y} = \sum_{i=1}^{r} d_i g_i y_i = \sum_{i=1}^{r} w_i y_i \qquad \text{where } r \text{ is the number of objects in the } sample \text{ that responded in a particular cell} \qquad (1)$$

The weights d_i are the original weights before calibration, and the weights $d_i g_i = w_i$ are the weights *after* calibration using information about register totals of some auxiliary variables from statistical registers. With formula (1) weighted sums and weighted frequencies are calculated. When calculating mean values, the weighted sums are divided by weighted frequencies.

For *censuses* where data is collected by the statistical office, nonresponse occurs in the same way as with sample surveys. Using the methods described in, for instance, Särndal and Lundström (2005), it is also possible here to calculate the weights g_i to reduce the errors caused by nonresponse.

7.1.2 Estimation methods for register-based surveys

In this section, we will discuss two questions:

– Are there estimation methods for register-based surveys?
– Can good quality be achieved with estimation methods where weights are used?

Are there estimation methods for register-based surveys?
No special methods are currently used when register-based statistics are produced; instead, calculations and summations are made in the simplest possible way.

$$\hat{Y} = \sum_{i=1}^{R} y_i \qquad \text{where } R \text{ is the number of objects in the } register \text{ in a particular table cell} \qquad (2)$$

We interpret these seemingly simple calculations as estimates, the values of these estimates depending on the methods used when the register was created. The work carried out to create a statistical register is described Chapters 5 and 6. If this work is carried out in different ways, there will be different numerical values in the register-based statistics that are produced with the register. Choosing the methodology for the creation of a register means also choosing an estimation methodology.

With sample surveys, the methodology work focuses on *how* to carry out the summing up, i.e. how the weights d_i and g_i are to be decided. Methodology work within register-based statistics is instead focused on *what* is to be summed up, i.e. how to define the register population, how to define the objects in this population and how the register's variables are to be formed with the available data. The way a statistical register is created determines which estimates will be made with the register. So there are estimation methods also within register-based statistics.

Chapters 5 and 6 in this sense deal with estimation methods. These methods we name the *fundamental estimation methods* for register-based surveys and the estimation methods presented in Chapters 8 and 9 are called *supplementary estimation methods*.

Can good quality be achieved with estimation methods where weights are used?
Besides the estimation methods that are decided by the way the register is created, we will in Chapters 8 and 9, introduce weights w_i to solve some of the quality problems. The weights are calculated in different ways for different problems and, with these weights, it is possible to correct for different types of errors, i.e. that the register estimates are on an incorrect level.

In register-based surveys, the weights $d_i = 1$ for objects without missing values, and $d_i = 0$ for objects with missing values. Estimates are here made by using formula (3):

$$\hat{Y} = \sum_{i=1}^{R} d_i g_i y_i = \sum_{i=1}^{R} w_i y_i \qquad \text{where } R \text{ is the number of objects in the } register \text{ in a particular cell} \qquad (3)$$

With traditional methods, all $w_i = 1$, but in Chapters 8 and 9 other weights will be used. The types of errors we discuss in these chapters include: errors due to discarding information in multi-valued variables, errors due to undercoverage, errors due to item nonresponse or missing values and errors due to level shifts in time series. The methodology could be used for more kinds of errors.

7.2 REGISTER-BASED SURVEYS – FUNDAMENTAL ESTIMATION METHODS

According to Chapter 5, the work to create a statistical register can be divided into the following phases. In all of these phases statistical work is done which influence the estimates that will be produced from the register. The influences of phases 1 and 3 are indirect and will not be discussed here.

1. (Determining the research objectives).

2. The inventory phase: What sources are available when a register is created?

3. (The planning phase).

4. Contacts with data suppliers and the receipt of administrative data.

5. The integration phase can be divided into three parts:

   a. How should the existing sources be integrated so that the register will contain the required object set, the register population?

   b. What processing should be carried out to check and correct object definitions? Derived objects are formed in the new register.

   c. The variables in the administrative sources are checked and edited. What processing should be carried out to create the variables in the register? Derived variables are formed in the new register.

Assume that we want to estimate equalised disposable income (average disposable income per consumption unit) of the households in different regions during a certain year. The chart below (similar to Chart 1.4 in Chapter 1) illustrates the work done to create the Income and Taxation Register (I&T).

In the chart, six external sources are shown, but in fact there are about 30 administrative registers which are used. The *inventory work* (phase 2) which is necessary to find these sources, the work necessary to *communicate with each administrative authority* (phase 4) so that each administrative source is correctly understood and used by the team responsible for the I&T-register have a very strong impact on the final estimates.

The *editing work* (phase 4 and 5, mainly 5c) where first each source is edited and corrected and the final work with consistency editing of all sources together, are also very important for the estimates that will be produced with the final I&T-register.

Chart 7.1 Data sources and register processing for the Income and Taxation Register

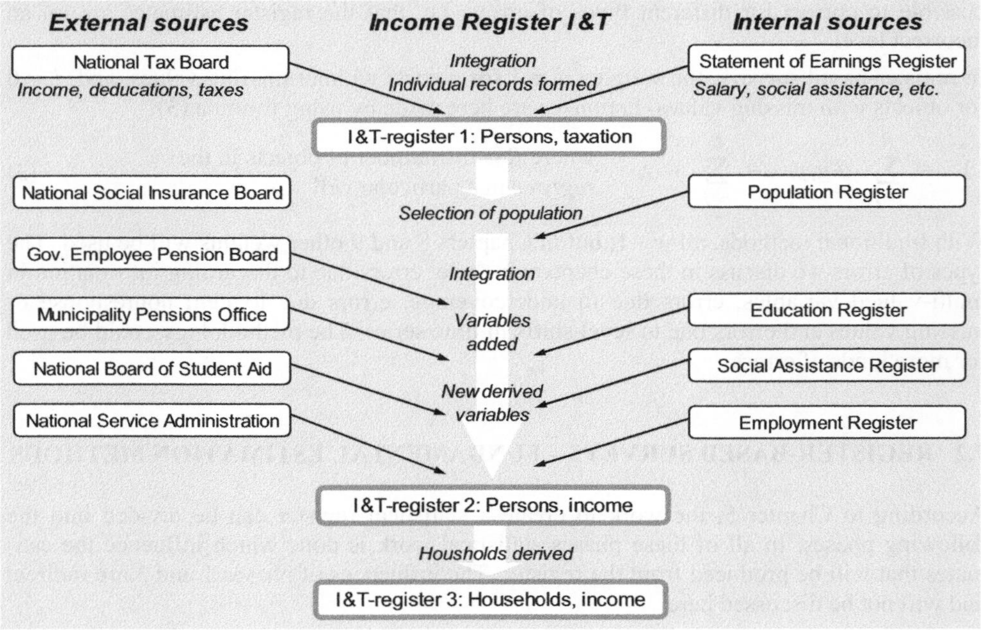

The way in which *the register population is defined* and created (phase 5a), is fundamental to the income estimates. If the population is defined as a calendar year population the income sum will be greater than if the population is defined as the population at the turn of the year. If population by region is defined according to where persons are administratively registered by the Tax Board or if actual addresses are used, will also influence the regional estimates.

The household unit in a register system is an object type that is derived with administrative information. The way *households are defined and created* (phase 5b) in the I&T-register is an essential part of the estimation.

Finally, derived variables can be created in different ways. The variable *equalised disposable income* is a derived variable, and the way this variable is created (phase 5c) is also a very important part of the estimation process.

7.3 USING WEIGHTS IN REGISTER-BASED SURVEYS

The fundamental estimation methods in Chapters 5 and 6 do not use weights, but the supplementary estimation methods in Chapters 8 and 9 will use different kinds of weights to reduce different kinds of errors or quality problems.

From a data matrix to statistical tables
How can weights be used in a register to make estimates? This is shown in the fictitious salary register in the chart below. The register contains data on monthly salaries for persons during a particular month and data on the extent of work *(Extent)*, where 1 is full-time.

The actual salary (in SEK, 8 SEK is approximately 1 USD or 1 Euro) is recalculated to a full-time salary using the variable *Extent*. The person's occupation is given with the occupational code *(ISCO)*. The occupational variable can be grouped into level of competence, where 1 is the lowest and 5 refers to managers. The variables age and full-time salary have also been divided into classes. To correct for any errors, such as nonresponse, the weights w_i have been calculated.

Chart 7.2 A salary register where the observations have weights

Person	Sex	Age	ISCO	Level	Salary	Extent	Salary Full	Salary Class	w_i	$w_i \cdot$ Salary$_i$	$w_i \cdot$ Extent$_i$	$w_i \cdot$ SalaryFull$_i$
(1)	(2)	(3)	(4)	(5)	(6)	(7)	(8)	(9)	(10)	(11)	(12)	(13)
PIN1	F	50–54	4190	2	14850	1.00	14850	14–14.9	1.028	15271.4	1.028	15271.4
PIN2	F	40–44	2330	4	16630	0.95	17505	17–17.9	1.031	17147.5	0.980	18049.8
PIN3	M	50–54	2492	4	17807	1.00	17807	17–17.9	1.083	19285.5	1.083	19285.5
PIN4	F	40–44	2330	4	1485	0.09	16500	16–16.9	1.031	1531.2	0.093	17013.5
PIN5	F	40–44	5133	2	6497	0.50	12994	12–12.9	1.031	6699.2	0.516	13398.4
PIN6	F	40–44	5131	2	14102	1.00	14102	14–14.9	1.031	14540.9	1.031	14540.9
PIN7	M	50–54	5131	2	858	0.06	14300	14–14.9	1.083	929.2	0.065	15487.3
...	...	...	...	...	...	...	...	...	...	...	...	...

The register contains columns (1)–(10) and, when the estimates are carried out, columns (11)–(13) are temporarily formed. The raw table below is formed by summing up the response variables in register columns (10)–(13) for all combinations of the spanning variables sex, age, ISCO (within this, also level of competence) and salary class.

The entire table in Chart 7.3 below, consists of around 2 200 rows. Register-based surveys are suitable for forming large detailed raw tables that can be used to create many different legible tables for different purposes.

Chart 7.3 A raw table that can be used to form many tables

Sex (1)	Age (2)	ISCO (3)	Level (4)	SalaryClass (5)	$\sum w_i$ (6)	$\sum w_i \cdot Salary_i$ (7)	$\sum w_i \cdot Extent_i$ (8)	$\sum w_i \cdot SalaryFull_i$ (9)
F	17–24	2330	4	12–12.9	42.52	429170	34.55	526165
F	17–24	2330	4	13–13.9	95.67	1293410	95.35	1297704
F	17–24	2330	4	14–14.9	42.52	201399	14.14	622852
F	17–24	2330	4	15–15.9	10.63	159444	10.63	159444
F	17–24	2330	4	16–15.9	53.15	163111	9.89	876942
...	...	...	...	...	...	...	...	...
M	60–64	8320	2	13–13.9	21.24	290107	21.23	290107
M	60–64	8320	2	14–14.9	10.62	149300	10.62	149300
M	60–64	9140	1	12–12.9	10.62	136422	10.62	136422
M	60–64	9140	1	13–13.9	10.62	71348	5.31	142697
M	60–64	9140	1	14–14.9	21.24	308040	21.23	308040

By further aggregating this raw table in different ways, more tables can be formed for different purposes. The variable salary is used both as a spanning variable (column 5) and as a response variable (columns 7 and 9) in the raw table above. In an actual statistical table, salary is only used in one of these two roles.

Retrieval of frequency tables – salary distribution for different study domains

By summing up column (6) in Chart 7.3, which contains the weighted number of persons, for different combinations of competences level, sex and salary class, the table in Chart 7.4 below is produced. The absolute frequencies in Chart A are recalculated to relative frequencies in the table in Chart B, illustrating the relation between, firstly, sex and salary and, secondly, between level of competences and salary.

Chart 7.4 Salary distribution by level and sex
A. Number of persons

Level:	1		2		3		4		5	
Salary	Wom.	Men	Wom.	Men	Wom.	Men	Wom.	Men	Wom.	Men
9–11.9	615	107	1823	484	83	32	31			
12–12.9	1138	108	2806	434	199	32	73			
13–13.9	2220	381	10382	1686	239	130	397	174		
14–14.9	560	162	9675	968	900	347	831	194		
15–15.9	114	54	4246	565	1719	533	911	228	21	
16–15.9	21	44	1709	651	1758	467	1293	454	10	
17–17.9			1389	520	1054	468	1675	576	124	
18–18.9	10	11	765	251	786	271	1729	721	114	11
19–19.9			196	122	487	229	1076	644	21	11
20–20.9			73	43	289	110	1492	882	31	21
21–22.9			21	22	237	66	550	567	62	44
23–25.9				11		22	238	412	238	250
26–29.9					10	11	114	205	186	163
30–34.9							52	151	10	99
35–39.9							155	230		44
40–125							145	492		33
Total	4677	869	33084	5758	7762	2717	10763	5930	817	675

B. Per cent

	1		...	5	
	Wom.	Men		Wom.	Men
9–11.9	**13.2**	12.3	...	**0.0**	0.0
12–12.9	**24.3**	12.5	...	**0.0**	0.0
13–13.9	**47.5**	43.8	...	**0.0**	0.0
14–14.9	12.0	**18.7**	...	**0.0**	0.0
15–15.9	2.4	**6.3**	...	**2.5**	0.0
16–15.9	0.4	**5.1**	...	**1.3**	0.0
17–17.9	0.0	0.0	...	**15.2**	0.0
18–18.9	0.2	**1.3**	...	**13.9**	1.6
19–19.9	0.0	**0.0**	...	**2.5**	1.6
20–20.9	0.0	**0.0**	...	**3.8**	3.2
21–22.9	0.0	**0.0**	...	**7.6**	6.4
23–25.9	0.0	**0.0**	...	**29.1**	**37.1**
26–29.9	0.0	**0.0**	...	**22.7**	**24.2**
30–34.9	0.0	**0.0**	...	**1.3**	**14.6**
35–39.9	0.0	**0.0**	...	0.0	**6.5**
40–125	0.0	**0.0**	...	0.0	**4.9**
Total	100.0	100.0	...	100.0	100.0

Retrieval of tables with means and ratios for different domains of study

By summing up the response variables in columns (6)–(9) in the raw table in Chart 7.3 for different combinations of the spanning variables in columns (1)–(4), tables with totals that can be used to form mean values and ratios for different domains of study are formed.

Chart 7.5 Table forming the basis for several different tables

Sex	Level	$\sum w_i$	$\sum w_i \cdot$ salary$_i$	$\sum w_i \cdot$ extent$_i$	$\sum w_i \cdot$ salaryfull$_i$	(6)/(3)	(4)/(5)	(5)/(3)
(1)	(2)	(3)	(4)	(5)	(6)	(7)	(8)	(9)
Wom.	1	4677	40635041	3061	61701517	13 191	13 274	0.65
Wom.	2	33084	360320838	25085	474211547	14 334	14 364	0.76
Wom.	3	7762	106558725	6416	129063365	16 627	16 607	0.83
Wom.	4	10763	175129111	9313	202168595	18 784	18 805	0.87
Wom.	5	817	17415674	771	18333690	22 431	22 583	0.94
Men	1	869	10905658	797	11795151	13 579	13 690	0.92
Men	2	5758	73156416	4898	85083524	14 777	14 936	0.85
Men	3	2717	42580548	2532	45632570	16 796	16 815	0.93
Men	4	5930	125232247	5430	136222043	22 971	23 063	0.92
Men	5	675	18050648	655	18594809	27 531	27 564	0.97

Chart 7.6

Mean salary by sex and level of compe-tences

Level	Women	Men
1	13 274	13 690
2	14 364	14 936
3	16 607	16 815
4	18 805	23 063
5	22 583	27 564

The table above is formed using column (8) in the table in Chart 7.5

Column (3) in the table in Chart 7.5 contains the weighted number of persons within each domain of study. Column (4) contains the weighted actual salaries, column (5) the weighted full-time jobs and column (6) the weighted full-time salaries. Column (7) contains the average full-time salary per person, column (8) the average salary per full-time job and column (9) the average extent of work per person.

The table in Chart 7.5 can be used to create a number of tables. The tables should have the layout of the table in Chart 7.6 where in the first place mean salaries for women and men can be compared.

The table in Chart 7.7 below shows parts of a table in which the average salaries for women and men are compared within the same age category and occupation. The full table contains around 800 rows and it is therefore appropriate to summarise the table's content by calculating standardised mean salaries. Register-based surveys are suitable for detailed table analyses that often should be supplemented with the calculation of standardised means with standard weights.

Women and men are differently distributed by age and occupation, which explains the largest part of the salary difference, 15 680– 18 860. If both women and men had the same distribution by age and occupation, according to the standard weights in column (8), the salary difference would only have been 16 256– 16 505 according to the two last columns.

Chart 7.7 Calculation of standardised mean salaries

Age	ISCO	Average salary, women	Average salary, men	Number of women	Number of men	Total number	Standard weighting	Women:	Men:
(1)	(2)	(3)	(4)	(5)	(6)	(7)	(8) = (7) / 5 688	(3) · (8)	(4) · (8)
17–24	2330	13 660	14 100	276	75	351	0.0062	84.41	87.13
...	...	...	...	...	...	...	...	...	...
60–64	7130	13 826	13 900	10	63	74	0.0013	18.09	18.19
Total		**15 680**	**18 860**	4 523	1 165	5 688	1.0000	**16 256**	**16 505**

7.4   ESTIMATION USING WEIGHTS – CALENDAR YEAR REGISTERS

In this section we discuss simple examples where weights should be used to produce estimates for register-based statistics. The calendar year register is the register version containing all objects that have existed at any point during a specific year. In a calendar year register, objects can exist during different time-periods. Most objects cause no problems, they exit during the whole year, but other objects are born or enter the register at certain moments and some other objects disappear or die during the calendar year. This gives rise to estimation problems that can be solved by using weights. Time can be used as a *weight-generating variable*, and with these weights correct estimates can be produced for calendar year registers.

Average population

The average population in a municipality can be estimated in the following way, where we calculate the day of birth or arrival in the municipality as a full day and the day when the person moved/died as no day in the municipality.

Chart 7.8 Calendar year register for the population in a (small) municipality

Person	Existed 1/1 2005	Arrived during 2005 yyyymmdd	Ceased during 2005 yyyymmdd	Existed 31/12 2005	Weight = Time in the municipality, years
PIN1	Yes	-	20050517	No	136/365 = 0.37
PIN2	Yes	-	-	Yes	365/365 = 1.00
PIN3	No	20050315	20050925	No	194/365 = 0.53
PIN4	No	20050606	-	Yes	209/365 = 0.57
Total	2			2	2.47

The traditional way of calculating the average population for 2005 is to form the average value of the population on 1/1 in 2005 (2 persons) and the population on 31/12 in 2005 (also 2 persons). A more specific calculation, in which time in the municipality is used as weight, gives the average population during 2005 as 2.47 persons instead of the traditional measure of 2.

Flow and stock variables

In the example below, the data relates to enterprises in a particular region during 2004. Certain enterprises began or ceased to exist at different times during the year, and time can also be used here as a weight-generating variable.

Flow and stock variables should be treated differently. Flow variables, such as value added of an enterprise, only relates to the values during the period of the year in which the enterprise was active, and therefore does not need to be weighted. A stock variable showing the level at a point in time, such as number of employees, must be weighted. The total value added in the region during 2004 was SEK 83 million, while the average number of employees was 112.5. Productivity is calculated as 83/112.5 = SEK 0.738 million per employee and year.

Chart 7.9 Calendar year register for 2004 for enterprises in a particular (small) region

Enterprise identity	Existed 1/1	Arrived	Ceased	Existed 31/12	Weight	Value added	Nr. of employees	Weight • Nr. empl.
EU1	Yes	-	20040630	No	0.50	10	30	0.50 • 30 = 15.0
EU2	Yes	-	-	Yes	1.00	42	45	1.00 • 45 = 45.0
EU3	No	20040401		Yes	0.75	31	70	0.75 • 70 = 52.5
Total					2.25	83		112.5

7.5   CALIBRATION OF WEIGHTS IN REGISTER-BASED SURVEYS

In Section 7.1.2 it is stated that a general way to estimate register-based statistics is to use formula (3) below:

$$\hat{Y} = \sum_{i=1}^{R} d_i g_i y_i = \sum_{i=1}^{R} w_i y_i \qquad \text{where } R \text{ is the number of objects in the register in a particular cell} \qquad (3)$$

In this section we illustrate how weights can be calibrated by an example based on the register in Chart 7.10 below. Of the 19 observations in the register, two have missing values, observations number 6 and 15. Four persons are not employed, and have therefore no Industry code, but these are not missing values.

Chart 7.10 Register on persons from two small regions

(1) PIN	(2) Sex	(3) District	(4) Employed	(5) Industry	(6) Education	(7) d_i	x_{1i} Sex=F	x_{2i} Sex=M	x_{3i} District=1	x_{4i} Employed=1	w_i
1	F	1	0	null	Low	1	1	0	1	1	0.98276
2	M	1	1	A	Low	1	0	1	1	1	1.15517
3	F	1	1	A	Low	1	1	0	1	1	1.13793
4	M	1	1	A	Medium	1	0	1	1	1	1.15517
5	F	1	1	A	Medium	1	1	0	1	1	1.13793
6	M	1	1	*Missing*	Low	0	0	1	1	1	0.00000
7	F	1	1	D	Medium	1	1	0	1	1	1.13793
8	M	1	1	D	High	1	0	1	1	1	1.15517
9	F	1	1	D	Medium	1	1	0	1	1	1.13793
10	M	1	0	null	Medium	1	0	1	1	1	1.00000
11	F	2	0	null	Low	1	1	0	0	0	1.00000
12	M	2	1	D	Low	1	0	1	0	0	1.17241
13	F	2	1	D	Low	1	1	0	0	0	1.15517
14	M	2	1	D	Medium	1	0	1	0	0	1.17241
15	F	2	1	D	*Missing*	0	1	0	0	0	0.00000
16	M	2	1	A	Low	1	0	1	0	0	1.17241
17	F	2	1	A	Medium	1	1	0	0	0	1.15517
18	F	2	1	A	Medium	1	1	0	0	0	1.15517
19	M	2	0	null	Medium	1	0	1	0	0	1.01724

If we want to estimate a frequency table describing education by industry with this register, the missing values will affect the estimates. The table in Chart 7.11 is based on the shaded columns in Chart 7.10 and simple summations with the weights d_i.

Chart 7.11 Persons by Education and Industry

	Industry A Number of persons	Industry D Number of persons	Industry A Per cent	Industry D Per cent
High education	0	1	0.0%	16.7%
Medium education	4	3	57.1%	50.0%
Low education	3	2	42.9%	33.3%
All	7	6	100.0%	100.0%

The variables in columns (2), (3) and (4) have no missing values, and these variables can be used to calibrate the weights d_i so that estimates using the calibrated weights w_i will be adjusted for the missing values in columns (5) and (6).

Sums and/or frequencies based on the variables without missing values can be used as calibration conditions. There are many ways to choose these, and each choice will give calibrated weights that can differ. In this example we will use four conditions: The correct number of women = 10, of men = 9, of persons in district 1 = 10 and of employed = 15.

If these four frequencies are estimated with the set of observations with missing values, the weights d_i should be used and due to missing values the estimates of the same statistics will be erroneous: The number of women = 9 (error = −1), of men = 8 (error = −1), of persons in district 1 = 9 (error = −1) and of employed = 13 (error = −2).

The idea with calibration is to adjust the weights d_i so that the errors of these four estimates will be zero. All other estimates will also be adjusted in the same manner and using the new weights it is possible to produce consistent estimates that have been adjusted for the missing values in the register.

The first seven columns in Chart 7.10 show the original register, while columns $x_{1i} - x_{4i}$ contain the information to be used when calibrating. In the calculations, x_i' vectors are used, one vector per row. For $i=1$, such as for $PIN1$, $x_1' = (1\ 0\ 1\ 0)$

The summations are now referring to all observations in the register, not only one cell as in the earlier formulas (1)–(3). The last column in Chart 7.10 shows the adjusted weights w_i, calculated in three steps:

1. $T = \Sigma d_i x_i x'_i$ and T^{-1} are calculated, where all $d_i = 1$ (missing values, $d_i = 0$) and $i = 1, 2, \ldots, 19$

    T is a matrix with squared and product totals, here a 4×4 matrix

    $$T = \begin{bmatrix} \Sigma d_i x^2_{1i} & \Sigma d_i x_{1i} x_{2i} & \Sigma d_i x_{1i} x_{3i} & \Sigma d_i x_{1i} x_{4i} \\ \Sigma d_i x_{2i} x_{1i} & \Sigma d_i x^2_{2i} & \Sigma d_i x_{2i} x_{3i} & \Sigma d_i x_{2i} x_{4i} \\ \Sigma d_i x_{3i} x_{1i} & \Sigma d_i x_{3i} x_{2i} & \Sigma d_i x^2_{3i} & \Sigma d_i x_{3i} x_{4i} \\ \Sigma d_i x_{4i} x_{1i} & \Sigma d_i x_{4i} x_{2i} & \Sigma d_i x_{4i} x_{3i} & \Sigma d_i x^2_{4i} \end{bmatrix}$$

2. The vector λ is calculated: $\lambda = T^{-1} (t_x - \Sigma d_i x_i)$

    The vector t_x is the 4 conditions for the number of women and men, persons in district 1, and persons employed.

    The vector $\Sigma d_i x_i$ is the corresponding unadjusted number.

t_x	$\Sigma d_i x_i$	$t_x - \Sigma d_i x_i$
10	9	1
9	8	1
10	9	1
15	13	2

The vector t_x represents the correct values of the four calibration conditions, and the vector $\Sigma d_i x_i$ represents the erroneous values based on the observations with missing values.

3. The adjusted weights become: $w_i = d_i (1 + x'_i \lambda)$. The adjusted weights are used to calculate weighted numbers and totals.

These formulas are illustrated below, where the calculations are done step by step.

1. The matrices T and T^{-1} are calculated:

$$T = \begin{bmatrix} 9 & 0 & 5 & 7 \\ 0 & 8 & 4 & 6 \\ 5 & 4 & 9 & 7 \\ 7 & 6 & 7 & 13 \end{bmatrix}$$

$$T^{-1} = \begin{bmatrix} 0.375000 & 0.250000 & -0.125000 & -0.250000 \\ 0.250000 & 0.362069 & -0.112069 & -0.241379 \\ -0.125000 & -0.112069 & 0.237069 & -0.008621 \\ -0.250000 & -0.241379 & -0.008621 & 0.327586 \end{bmatrix}$$

2. The vector λ is calculated:

$$\lambda = \begin{bmatrix} 0.375000 & 0.250000 & -0.125000 & -0.250000 \\ 0.250000 & 0.362069 & -0.112069 & -0.241379 \\ -0.125000 & -0.112069 & 0.237069 & -0.008621 \\ -0.250000 & -0.241379 & -0.008621 & 0.327586 \end{bmatrix} \bullet \begin{bmatrix} 1 \\ 1 \\ 1 \\ 2 \end{bmatrix}$$

$$\lambda = \begin{bmatrix} 0.000000 \\ 0.017241 \\ -0.017241 \\ 0.155172 \end{bmatrix}$$

3. The adjusted weights become: $w_i = d_i \, (1 + x'_i \lambda)$

For the first person in the register, $i=1$, and $x_1' = (1 \ 0 \ 1 \ 0)$

$$x'_1 \lambda = [1 \ 0 \ 1 \ 0] \bullet \begin{bmatrix} 0.000000 \\ 0.017241 \\ -0.017241 \\ 0.155172 \end{bmatrix} = -0.017241$$

The adjusted weight for person 1 becomes: $w_1 = 1 \bullet (1 - 0.017241) = 0.982759$

With adjusted weights the weighted frequencies in the table below are estimated. The relative frequencies happen to be almost the same as in Chart 7.11, but the number of persons now sum up to 15 (8.1 + 6.9) instead of 13.

Chart 7.12 Persons by Education and Industry, adjusted for missing values

	Industry A, weighted number of persons	Industry D, weighted number of persons	Industry A Per cent	Industry D Per cent
High education	0.0	1.2	0.0%	16.7%
Medium education	4.6	3.4	57.0%	49.7%
Low education	3.5	2.3	43.0%	33.6%
All	8.1	6.9	100.0%	100.0%

CHAPTER 8

Calibration and Imputation

Three issues are discussed in this chapter: missing values, overcoverage and level shifts in time series. Weights and the calibration of weights can be used as supplementary estimation methods in these three cases, and imputation can be used to adjust for missing values.

8.1   THE NONRESPONSE PROBLEM

Nonresponse adjustments are today common for sample surveys but not for register-based surveys. The tables below contain two examples from Statistics Sweden, and illustrate the two ways of publishing statistics with nonresponse.

The Labour Force Survey 2001		
Labour force category	000s	% of pop.
Employed	4 239	75.3
Unemployed	175	3.1
Not in labour force	1 218	21.6
Population aged 16–64	5 632	100.0

Note: The nonresponse rate in the Labour Force Survey is approximately 15%. The published estimates have been adjusted for the nonresponse.

Education Register 2001		
Educational level	000s	% of pop.
Less than 9 yrs	755	11.8
Comp. school 9 yrs	939	14.7
Upper secondary 2 yrs	1 747	27.4
Upper secondary 3 yrs	1 142	17.9
University < 3 yrs	802	12.6
University ≥3 yrs	848	13.3
Postgraduate	48	0.7
Education unknown	**106**	**1.7**
Population aged 16–74	6 386	100.0

Missing values in registers can be treated in different ways. One possibility, which is quite common, is to publish tables with one category *'Value unknown'*, and not to adjust for missing values at all. A second possibility is to use weights, which have been calibrated to reduce the effects of the missing values. A third possibility is to impute values when values are missing. We discuss these three possibilities, and at the end of this section present some conclusions on how missing values in a register should be treated. Also, in register-based surveys there are possibilities to reduce (item) nonresponse; this should always be done before adjustment methods are used. This reduction is done when the register is created, all possible sources should be utilised so that item nonresponse is minimised.

The methods used to adjust for nonresponse in sample surveys are discussed in many books and papers; however, missing values in register systems and in register-based surveys are topics that not have been widely investigated up to now. The important coordination and consistency between registers in the system adds to the requirements on the adjustment

Register-based Statistics – Administrative Data for Statistical Purposes A. Wallgren and B. Wallgren
© 2007 John Wiley & Sons, Ltd

methods, and this aspect is generally not treated in the literature on adjustment for sample surveys. The adjustment methods for registers should start with these methods, but further development is necessary.

Our aim here is not to discuss nonresponse adjustment generally, but only to discuss some problems specific to registers in a system. As a rule, we use the simplest possible application of each method, as we believe that most people working with statistical registers are not used to nonresponse adjustments. Adjustments based on weights and the calibration of weights are discussed by Särndal and Lundström (2005). An introduction and overview to imputation is given by Eltinge et al. (2003) and Pannekoek and de Waal (2005).

8.1.1   Missing values in registers

There are two nonresponse terms in surveys with their *own* data collection: *object nonresponse* (i.e. no response is received from certain objects), and *item nonresponse* (i.e. occasional questions have not been answered, but there are responses for the other questions). The concept of nonresponse is designed for surveys with their own data collection.

With register-based studies, it is difficult to differentiate between the concepts object nonresponse and undercoverage. The nonresponse term that is best suited to register-based surveys is the term *item nonresponse,* as an indication that certain values are missing. For *register-based surveys*, missing values can arise for different reasons:

− Variable values can be missing for certain objects due to flaws in the administrative system.

− With register-based surveys, registers are often created with the help of *several* source registers. When different source registers are linked and matched, mismatches due to missing objects in some source registers can occur. This results in missing values for all the variables that are imported from these registers.

− During the editing work (discussed in Section 6.3), missing values are discovered, or it is decided that certain values must be rejected.

For variables in a statistical register, the extent of the item nonresponse should always be reported. The methods for nonresponse adjustment will be described here. If imputing variable values, these should be shown in special variables so that it is always clear which are measured values and which are imputed values.

When a register is created, variables are imported from different source registers. If several of these imported variables contain item nonresponse, the new register will contain item nonresponse to a greater extent than the sources. There are certain qualitative variables that are used in many registers in which nonresponse must be treated in a consistent way, because of the stringent demands for consistency.

8.1.2   Make no adjustments, publish 'Value unknown'

In this section we give some examples where no adjustments for item nonresponse have been done. Our aim is to show the shortcomings of this way of publishing statistics.

Comparing small areas
The Swedish Employment Register contains data regarding gainfully employed persons, with industrial classification for the local unit where a person is employed, and the person's highest level of completed education. These variables are imported from the Business Register and the Education Register. Both sources contain item nonresponse or missing values.

According to the data in Chart 8.1, missing values regarding educational level is 1.7% and missing values regarding Industry is 79/5647 = 1.4%. Of the population in the Employment Register regarding the entire population aged 16–64, (79 + 93 – 2)/5647 = 3% lack values for industrial classification *and/or* educational level.

Chart 8.1
Population aged 16–64 by educational level and industrial classification 2001

000s persons	Employed within Industry …				Not empl	Total pop.	Employed within Industry …				Not empl	Total pop.
	A–F prod of goods	G–K private services	L–Q public services	Industry un-known			A–F	G–K	L–Q	Industry unknown		
							%	%	%	%	%	%
<9 yrs, Comp. school 9 yrs	273	273	141	17	611	1315	24.7	18.9	9.7	21.8	39.3	23.3
Upper secondary 2 yrs	403	404	416	22	324	1570	36.6	27.9	28.5	28.5	20.8	27.8
Upper secondary 3 yrs	229	369	217	15	252	1081	20.7	25.5	14.9	18.7	16.2	19.1
University < 3 yrs	107	188	273	11	177	757	9.7	13.0	18.7	13.6	11.4	13.4
University ≥ 3 yrs	86	204	410	12	119	830	7.8	14.1	28.0	15.4	7.7	14.7
Education unknown	5	10	4	**2**	73	**93**	0.4	0.7	0.3	2.0	4.7	1.7
Entire pop. aged 16–64	1103	1448	1462	**79**	1556	5647	100	100	100	100	100	100

The table above represents the way in which register-based statistics with missing values have traditionally been reported by Statistics Sweden. There are several different reasons for this: the missing value rate is considered small, and it has been judged that it will be too complicated to adjust. Our opinion is that the missing value rate is not small at all; that it is quite possible to adjust and that it is the responsibility of the statistical office to adjust for the effects of missing values.

In the two-way table above, there are missing values in both spanning variables. Even if the total missing value rate only amounts to 3%, this results in a table that is difficult to interpret. The patterns are disturbed by missing values in the table. By adjusting for missing values, tables will be easier to interpret for the users of the statistics.

However, the size of this missing value rate varies when different *municipalities* or other small categories are compared. The table below shows the highest and lowest shares of missing values for the country's municipalities.

Chart 8.2 Employment Register 2000 and 2001 – lowest and highest values of item nonresponse rates for population aged 16–64 in Sweden's municipalities

Item nonresponse:	Lowest 2000	Highest 2000	Lowest 2001	Highest 2001
Education unknown	0.3%	3.7%	0.4%	4.6%
Industry unknown	0.5%	6.0%	0.4%	3.7%
Both Education and Industry unknown	1.0%	7.0%	1.0%	7.5%

Even if missing value rate is not so large on a national level, it can be significant at municipality level, and it can also vary between years. Chart 8.2 shows that this is true, which means that the item nonresponse makes comparisons between municipalities and other small categories difficult. Many users will forget the category 'Industry unknown' when they compare different municipalities with regard to, for example, the size of the service sector.

How can a municipality with 1% missing value rate be compared to another municipality where the missing value rate was 7%? To make such comparisons, adjustments must be made for missing values, or the level estimates regarding different municipalities will not be comparable. In Chart 8.3 below, we have adjusted for missing values shown in Chart 8.1. Those with 'education unknown' were proportionately allocated across the educational levels in the same column, and then those with 'Industry unknown' were proportionately allocated across the three Industries on the same row.

Chart 8.3 Population aged 16–64 by educational level and industrial classification 2001 Estimates adjusted for missing values

000s persons	Employed within Industry			Not empl	Total pop.	Employed within Industry			Not empl	Total pop.
	A–F prod of goods	G–K private services	L–Q public services			A–F %	G–K %	L–Q %	%	%
<9 yrs, Comp.school 9 yrs	281	282	145	641	1337	25.0	19.1	9.8	41.2	23.7
Upper secondary 2 yrs	413	415	425	340	1597	35.7	28.1	28.6	21.8	28.3
Upper secondary 3 yrs	234	378	222	264	1100	20.8	25.6	14.9	17.0	19.5
University < 3 yrs	110	193	279	186	770	9.8	13.1	18.8	12.0	13.6
University ≥ 3 yrs	88	209	418	125	844	7.8	14.1	28.1	8.0	14.9
Entire population aged 16–64	1125	1476	1490	1556	5647	100.0	100.0	100.0	100.0	100.0

The table adjusted for missing values is easier to interpret, and the corresponding tables regarding different municipalities can also be compared. The argument that the missing value rate is small at an overall level does not justify the practice not to adjust, as the missing value rate can differ substantially between different small areas.

Item nonresponse varies over time

Attention should be paid to missing values within register-based statistics and it should be adjusted for in a way corresponding to that for sample surveys. For register-based surveys, it is common that missing values are *not* adjusted for, but is instead shown as a missing value. If the extent of the missing value rate varies over time, and if no corrections are made, the comparability over time will be of a low quality.

Example: Missing values in the Patient Register

The Patient Register kept by the Swedish National Board of Health and Welfare contains data on those who have received care in hospital. The diagnosis should be registered, but item nonresponse for this variable results in an underestimation of the number of patients with a particular diagnosis. Missing value rates can vary strongly between years and from region to region, depending on how well the administrative systems in different hospitals work. Chart 8.4 shows the difference between a time series that has *not* been adjusted for item nonresponse in the variable diagnosis and a time series that has been adjusted.

Chart 8.4 Falling accidents among boys aged 0–12 years in Norrbotten County
Number of accidents per 1 000 boys and year

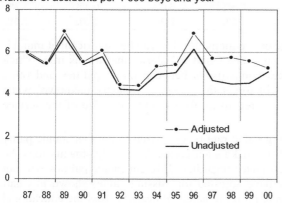

— Adjusted
—— Unadjusted

The time series pattern becomes incorrect when the missing value rate varies over time.

Comparisons of uncorrected values for different regions are misleading if the regions have different missing value rates.

Example: Missing values in the Swedish Education Register

Missing value rates in the Education Register have varied greatly over time, these variations producing apparent changes in the various time series, as seen in the Chart 8.5 below. To give the users of this register-based statistics a correct picture of the time series patterns, the effects of the missing values should be adjusted for. Between 1989 and 1990, missing value rates decreased from 5.7% to 1.4%. This was due to the data that was collected for the Population Census 1990. All series apart from compulsory school and postgraduate education increased between 1989 and 1990, increases which are to a large extent apparent. There is also a time series level shift between 1999 and 2000 due to changed educational classifications and the addition of new sources. This is discussed in Section 8.3.

Chart 8.5 Effects of missing values on time series from the Education Register Population aged 16–74 by educational level 1985–2000

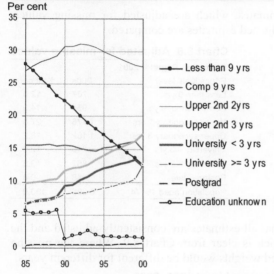

Per cent

8.1.3   Adjustment for missing values with weights

Which methods can be used to adjust for missing values? We show the simplest method, *straight expansion* for fictitious data from the Education Register. We have chosen the simplest adjustment method to illustrate the principle; a better adjustment could be done for the Education Register, if consideration was given to other variables such as age and sex, etc. But even this simple adjustment is better than none at all.

The weights in the register are adjusted or calibrated in accordance with the notation and methods given in Section 7.5. Straight expansion means that only one calibration condition is used: The total number of observations shall be the number of observations including those with missing values.

The adjustment is done for the Education Register in the chart below, where the original weight $d_i = 1$ (0 for the observations with missing value).

If the population consists of 6 386 015 persons, and the number of missing values is 106 051 and there are data for 6 279 964 persons, then the adjustment factor g_i would be 6 386 015/6 279 964 = 1.01689.

Chart 8.6 Adjustment for missing values in the Education Register 2001 with weights

Person	Sex	Age	Educational level	d_i	$d_i\, g_i = w_i$
PIN1	M	18	Compulsory school 9 yrs	1	1.01689
PIN2	F	72	Less than 9 yrs	1	1.01689
PIN3	M	33	Upper secondary 2 yrs	1	1.01689
PIN4	M	62	Upper secondary 3 yrs	1	1.01689
PIN5	F	71	**Missing value**	0	0
PIN6	F	26	University ≥ 3 yrs	1	1.01689
PIN7	M	54	Postgraduate	1	1.01689
PIN8	M	67	**Missing value**	0	0
PIN9	F	39	Less than 9 yrs	1	1.01689
...	...	...	...	...	...
PIN6386015	M	53	University < 3 yrs	1	1.01689
Total:				**6 279 964**	**6 386 015**

The weights w_i are used to calculate estimates, which are adjusted for missing values. Below, the unadjusted estimates and the adjusted estimates are compared.

Chart 8.7 Unadjusted table

Education Register 2001

Educational level	000s	% of pop.
Less than 9 yrs	755	11.8
Comp. school 9 yrs	939	14.7
Upper secondary 2 yrs	1 747	27.4
Upper secondary 3 yrs	1 142	17.9
University < 3 yrs	802	12.6
University ≥3 yrs	848	13.3
Postgraduate	48	0.7
Education unknown	**106**	**1.7**
Population aged 16–74	6 386	100.0

Chart 8.8 Adjusted for missing values

Education Register 2001

Educational level	000s	% of pop.
Less than 9 yrs	767	12.0
Comp. school 9 yrs	954	14.9
Upper secondary 2 yrs	1 776	27.8
Upper secondary 3 yrs	1 162	18.2
University < 3 yrs	816	12.8
University ≥ 3 yrs	862	13.5
Postgraduate	48	0.8
Population aged 16–74	6 386	100.0

Adjusted weights in a data matrix mean that all estimates are consistently adjusted and the comparability over time is improved, which is clear from Chart 8.9. Where the missing value rate varies between years, the adjusted weights would be different for different years.

Chart 8.9 Population aged 16–74 by educational level 1985–2000
A. Unadjusted series   B. Series adjusted for missing values

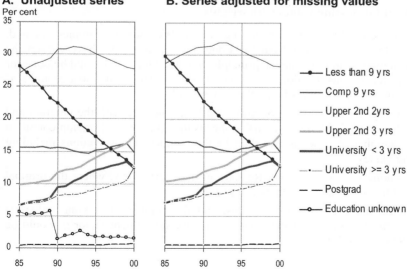

8.1.4 Adjustment for missing values with imputation

Another way to adjust for the effects of item nonresponse is to form *imputed values* when variable values are missing. The missing values are then replaced by synthetic values. There are two different ways of forming such values:

– The value is formed randomly using one or more probability distributions. This method applies to qualitative variables.

– The value is formed using a (deterministic) model in the same way as with derived variables, as described in Section 6.2.3.

The advantage with the imputation of variable values is that it avoids the need to calculate with weights, and the distribution of all other variables in the register remains unchanged.

Imputed values for qualitative variables formed randomly

The Chart 8.10 below shows how to form imputed values for the variable *educational level*. As we are dealing with a register on persons, the value for persons PIN5 and 8 should not be imputed according to Swedish law; instead, *synthetic observations* that do not have personal identification numbers should be formed.

The imputation, which corresponds to straight expansion, is carried out as follows: the observations with missing values on *educational level* are used to form the same number of synthetic observations. These synthetic observations get values for the variable *educational level* completely at random. These randomly chosen educational levels follow the same distribution as among those for which data on *educational level* is known. The register is increased with random numbers, and then a data matrix is created without personal identification numbers. These random numbers are values for a technical variable used internally.

Chart 8.10 Adjustment for missing values in the Education Register with imputation

A. Actual register 2001

	Sex	Age	Educational level	Random number
PIN1	M	18	Compulsory school 9 yrs	0.7771
PIN2	F	72	Less than 9 yrs	0.3168
PIN3	M	33	Upper secondary 2 yrs	0.3096
PIN4	M	62	Upper secondary 3 yrs	0.8667
PIN5	F	71	**Missing value**	**0.1749**
PIN6	F	26	University ≥ 3 yrs	0.4114
PIN7	M	54	Postgraduate	0.1605
PIN8	M	67	**Missing value**	**0.5536**
PIN9	F	39	Less than 9 yrs	0.5513
...	...	...	...	...
PIN6386015	M	53	University < 3 yrs	0.7828

B. Data matrix for analysis 2001

Sex	Age	Educational level	Educational level imputed
M	18	Compulsory school 9 yrs	No
F	72	Less than 9 yrs	No
M	33	Upper secondary 2 yrs	No
M	62	Upper secondary 3 yrs	No
F	71	**Compulsory school 9 yrs**	Yes
F	26	University ≥ 3 yrs	No
M	54	Postgraduate	No
M	67	**Upper secondary 3 yrs**	Yes
F	39	Less than 9 yrs	No
...	...	...	...
M	53	University < 3 yrs	No

C. Probability distribution based on frequency table in Chart 8.8

Educational level	Share of population	Accumulated share
Less than 9 yrs	0.120	0.120
Compulsory school 9 yrs	0.149	0.269
Upper secondary 2 yrs	0.278	0.547
Upper secondary 3 yrs	0.182	0.729
University < 3 yrs	0.128	0.857
University ≥ 3 yrs	0.135	0.992
Postgraduate	0.008	1.000
Population aged 16–74	**1.000**	

Random numbers in the register are uniformly distributed between 0 and 1. Persons with a random number between 0 and 0.120 are given the level *less than 9 years*, and those with a random number between 0.120 and 0.269 are given the level *Compulsory school 9 years*, etc.

By using the relationships between age, sex and educational level, the imputation can be improved. For different combinations of age category and sex, different frequency distributions for educational level are used. Chart 8.11 below compares three such distributions. There are significant differences between these, which means that it is possible to improve the adjustment for missing values by using different distributions when the values are randomly distributed for different combinations of sex and age.

Chart 8.11 Frequency table by age and sex, Education Register 2001

Educational level	Accumulated share Men aged 65–74	Accumulated share Women aged 65–74	Accumulated share Both aged 16–74
Less than 9 yrs	0.466	0.455	0.120
Comp. school 9 yrs	0.507	0.532	0.269
Upper secondary 2 yrs	0.700	0.821	0.547
Upper secondary 3 yrs	0.837	0.858	0.729
University < 3 yrs	0.901	0.918	0.857
University ≥ 3 yrs	0.988	0.998	0.992
Postgraduate	1.000	1.000	1.000

In the Chart 8.12 below, the same register is used with the same random numbers as previously. However, the random numbers have been translated here into educational level by using other frequency tables. Women in the age category 65–74 years with a random number between 0 and 0.455 are given the level *Less than 9 years*. Men in the age category 65–74 yrs with a random number between 0.507 and 0.700 are given the level *Upper secondary 2 years*. In the same way, younger persons with '*Missing value*' are given an imputed value using frequency tables for their age categories and sex.

Chart 8.12 Adjustment for missing values in the Education Register with imputation
A. Actual register 2001   **B. Data matrix for analysis 2001**

	Sex	Age	Educational level	Random number	Educ.level imputed	Sex	Age	Educational level
PIN1	M	18	Compulsory school 9 yrs	0.7771	No	M	18	Compulsory school 9 yrs
PIN2	F	72	Less than 9 yrs	0.3168	No	F	72	Less than 9 yrs
PIN3	M	33	Upper secondary 2 yrs	0.3096	No	M	33	Upper secondary 2 yrs
PIN4	M	62	Upper secondary 3 yrs	0.8667	No	M	62	Upper secondary 3 yrs
PIN5	F	71	**Missing value**	**0.1749**	Yes	F	71	**Less than 9 yrs**
PIN6	F	26	University ≥ 3 yrs	0.4114	No	F	26	University ≥ 3 yrs
PIN7	M	54	Postgraduate	0.1605	No	M	54	Postgraduate
PIN8	M	67	**Missing value**	**0.5536**	Yes	M	67	**Upper secondary 2 yrs**
PIN9	F	39	Less than 9 yrs	0.5513	No	F	39	Less than 9 yrs
...	...	...	...	...	...	...	...	...
PIN6386015	M	53	University < 3 yrs	0.7828	No	M	53	University < 3 yrs

PIN5, a 71 year old woman has the educational level *Less than 9 yrs*, which differs from the imputation in Chart 8.10, where she is given *Compulsory school 9 yrs*. The imputed level for *PIN8* is also changed to a shorter period of education.

When is it appropriate to use randomly imputed values?
The above method is appropriate when describing a qualitative variable with item nonresponse, possibly divided into different categories, such as age, sex and region. After a high-quality imputation, the levels are more comparable between categories and over time than if no adjustments are made for missing values.

If it is the relationship between a variable y and a qualitative variable x that is to be studied, where the x variable has item nonresponse, randomly imputed values for the x variable should not be used. For instance, it would not be appropriate to use randomly imputed educational levels when describing the average monthly salaries for different educational levels. In this case, it would be better to calculate the average salary only for the persons for whom the educational level is known.

Imputed values formed using a deterministic model

The imputation method used in Chart 8.12 above utilises the relationship between the x variables age and sex, and the y variable educational level. However, the imputed values are also formed *randomly*. For a particular combination of age and sex, educational level is not determined exactly but only randomly. We give some examples below of imputation methods where the values of the x variables determine the imputed values *exactly*. The models used for this type of imputation are called *deterministic models*.

Section 6.2.3 discusses how derived variables can be formed with deterministic causal models. Imputed variable values can be formed in a similar way, the difference being that derived variable values are calculated for *all* objects in the data matrix, while imputed variable values are only formed for those objects that have missing values due to item nonresponse.

When the editing work is carried out, it is detected that the value is missing or that certain variable values are unreasonable and must be rejected. This leads to the calculation of imputed values in close connection with the editing work. The editing case studies presented in Section 6.3 contain several examples of imputation methods.

When editing the Income and Taxation Register, it is discovered that social assistance has not been reported for some municipalities (see Section 6.3.1). For households in these municipalities, the previous year's values are therefore imputed. A simple model, *this year's assistance = previous year's assistance,* is used when imputing. On a household level, therefore, *modelling errors* or *imputing errors* can occur if the year's assistance differs from the previous year's. Attempts should be made to use models that give as small imputing errors as possible.

When the demands for quality are so high that imputing errors cannot be accepted, it is advisable to carry out a *special data collection*. The objects that lack values for an important variable can then provide the missing values via a questionnaire or an interview. Section 6.3.2 describes the editing of the annual statement of earnings data. For all statement of earnings data, a local unit identity should be given. When these data are missing or are considered unreasonable, the employer is contacted.

Editing of the enterprises income declarations gives examples of different types of imputation methods:

– Data on the number of full-year employees are taken from annual reports. If these data are missing, imputed values are formed by calculating an estimate of the number of full-year employees by dividing the enterprise's wage sum by the average wage per full-year employee in the industry. The average wage for the industry has first been calculated using those enterprises for which both the number of full-year employees and wage sums are known.

– The register population in Structural Business Statistics lacks economic variable values for some enterprises. Data on Industry and number of full-year employees have been imported from the Business Register. For enterprises where Industry, number of full-year employees and economic variables are known, tables are formed with the mean values for the different economic variables, by Industry and number of full-year employees. These tables are a form of model, which for given values of Industry and number of full-year employees shows how imputed values should be formed using the calculated mean values.

8.1.5   Missing values in a system of registers

When different registers are integrated and variables are imported from one register to other registers, quality flaws such as missing values are also imported to these other registers.

For example, the variable Industry is created in the Business Register and is then imported into other business registers, activity registers, registers on persons and also into real estate registers. This means that it is not sufficient to adjust for missing values in the Industry variable only in one register; the adjustment method must adjust for missing values in this variable in the whole register system in a consistent manner.

After trying to reduce the item nonresponse rate by using more sources, and perhaps also by collecting information from certain categories of objects, the estimates of register-based statistics should be adjusted for missing values. We compare here the two methods of adjusting: using weights or imputing values.

Adjustment for missing values with weights in a system of registers
The Population, Education and Employment Registers relate to the population on 31 December of a particular year. There is no item nonresponse in the Population Register; the Education Register contains item nonresponse in the variable educational level; and the Employment Register contains item nonresponse in both variables educational level and industrial classification.

If each register is adjusted separately for missing values with weights, the weights for the same person will be different in the three different registers. This is illustrated in Chart 8.13 below. Statistics from the three registers will then be inconsistent; for example, the number of 18-year-old men will be different (PIN1 has different weights in Chart 8.13 A, B and C).

If statistics from different registers that relate to the same population are to be consistent, weights must be calculated jointly, and the same weights must be used for all the registers. This can be difficult to achieve. Our conclusion is that adjustment for missing values with weights will cause problems for coordination and consistency within the register system.

Chart 8.13   Adjustment for missing values with weights in a system of registers

A. Population Reg.

Person	Sex	Age	d_i
PIN1	M	18	1
PIN2	F	72	1
PIN3	M	33	1
PIN4	M	62	1
PIN5	F	71	1
PIN6	F	26	1
PIN7	M	54	1
PIN8	M	67	1
PIN9	F	39	1
...	...	...	...

B. Education Register

PIN	Educ. level	$d_i g_i$
PIN1	Comp school 9 yrs	1.01689
PIN2	Less than 9 yrs	1.01689
PIN3	Upper 2nd 2 yrs	1.01689
PIN4	Upper 2nd 3 yrs	1.01689
PIN5	**Missing value**	0
PIN6	University ≥ 3 yrs	1.01689
PIN7	Postgraduate	1.01689
PIN8	**Missing value**	0
PIN9	Less than 9 yrs	1.01689
...	...	...

C. Employment Register 16–64 years

PIN	Industry	Educ. level	$d_i g_i$
PIN1	DM	Comp school 9 yrs	1.02930
-	-	-	-
PIN3	**Missing**	Upper 2nd 2 yrs	**0**
PIN4	DK	Upper 2nd 3 yrs	1.02183
-	-	-	-
PIN6	DB	University ≥ 3 yrs	1.02326
PIN7	DK	Postgraduate	1.02326
-	-	-	-
PIN9	DM	Less than 9 yrs	1.02930
...	...	...	...

Note: Three persons, PIN2, PIN5 and PIN8, are not gainfully employed according to the Employment Register, and are not between 16–64 years old. The weights $d_i g_i$ in Chart 8.13 B are the same as in Chart 8.6 and the weights $d_i g_i$ in C have been calculated by comparing the number of persons in different cells in Chart 8.1 and 8.3. For example 281/273=1.02930

Adjustment for missing values with imputation in a system of registers

If different registers in the system are adjusted for missing values using imputation in the way described in Section 8.1.4, the statistics from different registers could be completely consistent. At the same time as a variable is imported, the random numbers (or the imputed values) used in the original register are also imported. Imputations can then be done which are consistent between the different registers.

The example below shows how it is possible to import educational level from the Education Register and industrial classification from the Business Register to the Employment Register. Missing values in all three of these registers can then be replaced with the imputed values in a consistent way.

Chart 8.14 Adjustment for missing values in the Education Register using imputation

A. Actual register

Person	Sex	Age	Educational level	Random number
PIN1	M	18	Comp school 9 yrs	0.7771
PIN2	F	72	Less than 9 yrs	0.3168
PIN3	M	33	Upper 2nd 2 yrs	0.3096
PIN4	M	62	Upper 2nd 3 yrs	0.8667
PIN5	F	71	**Missing value**	**0.1749**
PIN6	F	26	University ≥ 3 yrs	0.4114
PIN7	M	54	Postgraduate	0.1605
PIN8	M	67	**Missing value**	**0.5536**
...	...	...	...	...

B. Data matrix for analysis

Sex	Age	Educational level	Educ. level imputed
M	18	Comp school 9 yrs	No
F	72	Less than 9 yrs	No
M	33	Upper 2nd 2 yrs	No
M	62	Upper 2nd 3 yrs	No
F	71	**Comp school 9 yrs**	Yes
F	26	University ≥ 3 yrs	No
M	54	Postgraduate	No
M	67	**Upper sec 3 yrs**	Yes
...	...	...	

Chart 8.15 Adjustment for missing values in the Business Register using imputation

A. Actual register

Enterprise	Industry	Random number
LeU1	DB	0.0316
LeU2	DK	0.6444
LeU3	**Missing value**	**0.3978**
LeU4	DA	0.2846
LeU5	DK	0.2044
...	...	...

B. Data matrix for analysis

Industry	Industry imputed
DB	No
DK	No
DM	Yes
DA	No
DK	No
...	...

Chart 8.16

Adjustment for missing values in the Employment Register with imputation

A. Actual register

Person	Enter-prise	Industry	Random number Industry	Educational level	Random number Education
PIN1	LeU5	DK	0.2044	Comp school 9 yrs	0.7771
PIN2	-	-	-	Less than 9 yrs	0.3168
PIN3	LeU3	**Missing**	**0.3978**	Upper 2nd 2 yrs	0.3096
PIN4	LeU2	DK	0.6444	Upper 2nd 3 yrs	0.8667
PIN5	-	-	-	**Missing value**	**0.1749**
PIN6	LeU1	DB	0.0316	University ≥ 3 yrs	0.4114
PIN7	LeU5	DK	0.2044	Postgraduate	0.1605
PIN8	-	-	-	**Missing value**	**0.5536**
...	...	...	...	...	...

B. Data matrix for analysis

Indu-stry	Industry imputed	Educational level	Educ. level imputed
DK	No	Comp school 9 yrs	No
-	-	Less than 9 yrs	No
DM	Yes	Upper 2nd 2 yrs	No
DK	No	Upper 2nd 3 yrs	No
-	-	**Comp school 9 yrs**	Yes
DB	No	University ≥ 3 yrs	No
DK	No	Postgraduate	No
-	-	**Upper 2nd 3 yrs**	Yes
...	...	...	...

Final conclusions

The conclusions to be drawn from this discussion are that adjustment for missing values should be done; adjustments must be coordinated; and that imputation is the most appropriate method for the adjustment of item nonresponse in a register system.

Within the system, the Education and the Business Registers are responsible for the adjustment for item nonresponse of *Education* and *Industry* respectively. Other registers should then use these adjustments.

8.2   ESTIMATION METHODS TO CORRECT FOR OVERCOVERAGE

In this section we show how calibration can be used to correct for overcoverage in a register. Currently, this is not usually corrected but, using the methods in this section, it could be possible to try out methods that can correct for these sources of error.

Overcoverage in the Population Register

The first sign that there is overcoverage in Statistics Sweden's Population Register came from demographic studies on mortality. Among a few categories of foreign-born persons, mortality was strangely low. Furthermore, it was found that the share of families, with no information on disposable income, was high among certain categories of immigrants.

Overcoverage in the Swedish Population Register has been estimated by Greijer (1995, 1996, 1997a, 1997b), who analysed nonresponse in the Labour Force Surveys and in a census on foreign-born persons based on a postal questionnaire. Using this information, it was possible to estimate overcoverage among different categories of foreign-born persons.

Data in statistical registers can also be used to give indications of overcoverage. A foreign-born person without income in any register can have left Sweden without reporting this to the tax authorities.

Overcoverage can cause serious errors in register-based statistics. For instance, the average income for those born in different countries can be misleading. For persons born in certain countries, the underestimation can be around 20%.

How should we control overcoverage and improve quality? The strategy for correcting errors caused by overcoverage can include the following:

1. By being watchful when carrying out macro editing, it is possible to find unreasonable estimates in the register-based statistics. The question should be asked, whether overcoverage could be the cause of these extreme estimates.

2. If overcoverage is suspected, available sample surveys and other sources can be used to help estimate this overcoverage.

3. Overcoverage can be estimated for different categories in the register once enough information on the extent and character of the overcoverage has been collected.

4. The weights can then be adjusted to correct for the estimated overcoverage. Before adjustment, all weights are equal to 1; after adjustment, the weights for the different categories for which there is overcoverage will be less than 1. Use *calibration methods* (Section 7.5) to adjust the weights when overcoverage is described by many variables.

5. The adjusted weights are stored in the base register (in this case the Population Register).

6. All other statistical products using the base register will then use the weights. In this way, all the statistics produced will be consistently corrected for the estimated effect of overcoverage.

Example: In a (fictitious) register with 1 000 foreign-born persons, overcoverage is as much as 10%, i.e. the register's 1 000 objects correspond to 900 persons in reality. We assume further that overcoverage has been estimated for different categories according to Chart 8.17 below, and that we use this information to calculate the adjusted weights.

Chart 8.17 Estimated overcoverage for different categories

		Number of persons before correction (1)	Estimated overcoverage (2)	Number of persons after correction for overcoverage (3)
Country of birth	Europe	*584*	6.7%	**545**
	Not Europe	*416*	14.7%	**355**
	Total	1000	10.0%	900
Years in Sweden	Few	*819*	7.2%	**760**
	Many	181	22.7%	140
	Total	1000	10.0%	900
Income	Low	101	40.6%	60
	High	*899*	6.6%	**840**
	Total	1000	10.0%	900

In Chart 8.17, it appears that we have six expressions for the number of persons but there are actually only four expressions because, with the four numbers, it is possible to calculate the remaining numbers. For example, using the four numbers marked in bold, it is possible to calculate the remaining numbers. These four numbers after correction and the corresponding numbers before correction (in italics) will be used as calibration conditions below.

Calibration of weights

Chart 8.18 shows how adjusted weights are calculated using the calibration methodology described in Section 7.5. The first five columns show the original register, while columns $x_{1i} - x_{4i}$ contain the information to be used when calibrating. In the calculations, $x_i{'}$ vectors are used, one vector per row. For $i=1$, such as for $PIN1$, $x_1{'} = (1 \ 0 \ 1 \ 1)$. The four calibration conditions define the vector t_x with the correct absolute frequencies and the vector $\Sigma d_i x_i$ with the incorrect absolute frequencies. These four conditions are found in Chart 8.17.

Chart 8.18 Register with calibrated weights   **Conditions**

PIN	Country of birth	Years in Sweden	Income	Weights d_k	x_{1i} Country Europe	x_{2i} Country Not Eur	x_{3i} Few years	x_{4i} High income	Adjusted weight, w_i	t_x	$\Sigma d_i x_i$
1	Europe	Few	High	1	1	0	1	1	0.992	545	584
2	Not Eur	Few	High	1	0	1	1	1	0.916	355	416
3	Europe	Few	Low	1	1	0	1	0	0.657	760	819
4	Europe	Few	High	1	1	0	1	1	0.992	840	899
5	Not Eur	Many	High	1	0	1	0	1	0,770		
6	Not Eur	Few	Low	1	0	1	1	0	0.581		
...	...	...	...	...	...	...	...	...	...		
1000	Not Eur	Few	High	1	0	1	1	1	0.916		
Total				**1 000**	**584**	**416**	**819**	**899**	**900**		

The overcoverage in the Swedish Population Register may seem to be of small importance – out of 9 million records only 50 000 records are suspected to be overcoverage, i.e. 0.6%. But many important ways of using the register are hampered by this error, e.g. the comparisons between different kinds of immigrants. The identities of these 50 000 persons are not known, but its is known that they belong to certain categories of persons. With weights in the Population Register for these categories, the overcoverage errors can be reduced.

8.3   METHODS TO CORRECT FOR LEVEL SHIFTS IN TIME SERIES

Statistical registers from different years are used to produce time series. Thousands of time series can be produced by one statistical register and, when *the administrative system* that is the source for the statistical register changes, this can give rise to level changes in many time series. Changes made by the statistics producer can also cause level changes. *Classification systems*, such as those for industrial classification, occupation or education, are occasionally revised and these changes also cause level changes. *Quality changes* due to revised methodologies or new sources can also lead to level shifts.

Chart 8.19 Three ways of reporting level shifts in time series

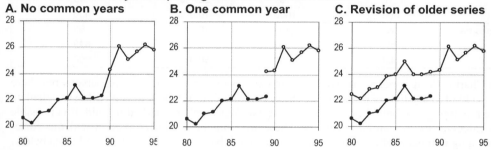

Chart 8.19A shows a situation in which the producer has not made any effort to measure the effect of a level shift; in Chart B, the producer has carried out double calculations for 1989 to illustrate the level change. In Chart C, the older time series for earlier years has also been recalculated.

Simply ignoring the level shift, as shown in the left chart should not occur – the users are then left to interpret completely by themselves. It is not sufficient only to mention that there is a level shift, it should also be explained how the data should then be interpreted. In publications, an observation such as *'Results from previous years should be interpreted with caution'* is sometimes used. However, no user will be satisfied with this as it is not clear in what way the data should be interpreted 'with caution'.

It should be a minimum requirement that the responsible producer carries out calculations as in Chart 8.19B, i.e. to show the effect on the year in which the change took place. But the best solution is shown in the Chart 8.19C, where time series are linked to produce longer series in which the effect of the level shift is minimised.

Correcting for level shifts in time series is called *linking time series*. We differentiate between *linking at macro level* and *linking at micro level*. It is common to link at macro level, i.e. working only with the aggregated time series values and trying to correct these for the effects of time series level shifts. This method has the disadvantage in that it can be difficult to ensure consistency among several linked series. For example, revised series for men and women must sum up to the revised series for the population as a whole. With many series, there are also many consistency requirements that have to be fulfilled and linking at the macro level becomes problematic.

Usually, those responsible for a statistical register can link a *limited number* of the most important time series when a change has taken place that affects the register. But what should be done with the possible thousands of other series that are also affected? Supposing that level changes have been estimated for 100 series but that the statistical register actually generates a total of 10 000 series. What can be done for the other 9 900 series?

The method of linking at micro level does not have these disadvantages, as all possible time series are linked at the same time and all these series will be consistent. Linking at micro level is carried out by calculating the revised weights in data matrices for the years or periods that are to be revised. The old time series values are given by the original weights, while the revised weights can be used to calculate the linked series.

We discuss below register processing that makes it possible to calculate revised time series values for earlier time periods. To be able to use the method, the effects of the change have first been estimated for the earlier time period by linking at the macro level for a limited number of series.

8.3.1 Estimating effects of level shifts – linking at macro level

If changes are made within the administrative system, the statistics producer should follow the changes that are made and measure the level changes. This can be done with information in the administrative registers, or by using sample surveys. Level shifts for the most important time series can then be estimated.

In 2000, a new educational classification system was introduced in the Swedish Education Register and, at the same time, new sources were used for the first time. At the Education Register the effects of the change 1999/2000 were analysed. The population aged 16–74 by different types of education, both according to the old classification with the old sources, and after the new classification and when new sources had been added, were compared. We have added three columns in Chart 8.20 where we have also adjusted for missing values.

Chart 8.20 Estimation of level shifts in 1999 for different kinds of education

Persons 16–74 years by level of education	Without adjustment for missing values:		After adjustment for missing values, using straight expansion:		
Educational level	**1999 old**	**1999 new**	**1999 old**	**1999 new**	**Difference**
Less than 9 yrs	863 332	848 289	877 948	862 428	−15 520
Comp school 9 yrs	1 025 322	926 996	1 042 680	942 446	−100 234
Upper secondary 2 yrs	1 768 378	1 798 085	1 798 316	1 828 054	29 738
Upper secondary 3 yrs	1 024 001	1 091 800	1 041 337	1 109 997	68 660
University < 3 yrs	846 876	748 071	861 213	760 539	−100 674
University ≥ 3 yrs	660 142	776 418	671 318	789 359	118 041
Postgraduate	41 619	41 619	42 324	42 313	−11
Education unknown	105 466	103 858	0	0	0
Population aged 16–74	**6 335 136**	**6 335 136**	**6 335 136**	**6 335 136**	**0**

How can this information now be used to correct the old time series values? A common method is the *ratio method*, which here involves the multiplication of all the old time series values with the ratio between new and old values for 1999. For the series *Persons with less than 9 yrs of education* the ratio is 862 428/877 948 = 0.982. Correspondingly, all the old time series values for the six other series could be multiplied by their corresponding ratios. However, this method of linking series gives insufficient consistency. The seven time series will not sum up to the total population for the years before 1999.

If the series are to be added, an additive model should be used, rather than a multiplicative model such as the ratio method. Chart 8.21 below shows one possible method of linking the series using an additive model. For 1999, the series are corrected according to the differences that we know apply for this year (Chart 8.20 above). We then assume that these differences have taken place gradually throughout the period 1990–1999, so we reduce the level differences by one tenth every year. These corrections are consistent, the total number of persons aged 16–74 does not change. The years before 1990 and after 1999 are not corrected. We illustrate the principles of the method with only these seven linked times series. An even better linking should be done in which attention is also paid to field of education, age, sex, etc. so that perhaps about 100 series first are linked at the macro level. The linked series will be used as calibration conditions in Section 8.3.2 below.

Chart 8.21 Additive time series corrections, number of persons

	Less than 9 yrs	Comp school 9 yrs	Upper2nd 2 yrs	Upper2nd 3 yrs	University < 3 yrs	University ≥3 yrs	Post-graduate	Total
1990	−1 552	−10 023	2 974	6 866	−10 067	11 804	−1	0
1991	−3 104	−20 047	5 948	13 732	−20 135	23 608	−2	0
1992	−4 656	−30 070	8 921	20 598	−30 202	35 412	−3	0
1993	−6 208	−40 094	11 895	27 464	−40 270	47 216	−4	0
1994	−7 760	−50 117	14 869	34 330	−50 337	59 020	−5	0
1995	−9 312	−60 140	17 843	41 196	−60 404	70 824	−7	0
1996	−10 864	−70 164	20 817	48 062	−70 472	82 629	−8	0
1997	−12 416	−80 187	23 790	54 928	−80 539	94 433	−9	0
1998	−13 968	−90 210	26 764	61 794	−90 607	106 237	−10	0
1999	**−15 520**	**−100 234**	**29 738**	**68 660**	**−100 674**	**118 041**	**−11**	**0**

Chart 8.22C below shows the effect of this time series linking. The largest time level shifts, for Compulsory school, Upper secondary 3 yrs and both University series, have disappeared. By comparing Charts 8.22 A, B and C, the effect of the different stages on the improvement of comparability over the years can be seen.

Chart 8.22 Population aged 16–74 by educational level 1985–2000
A. Unadjusted series   B. Adjusted for   C. Adjusted for missing values
                                    missing values   and linked series

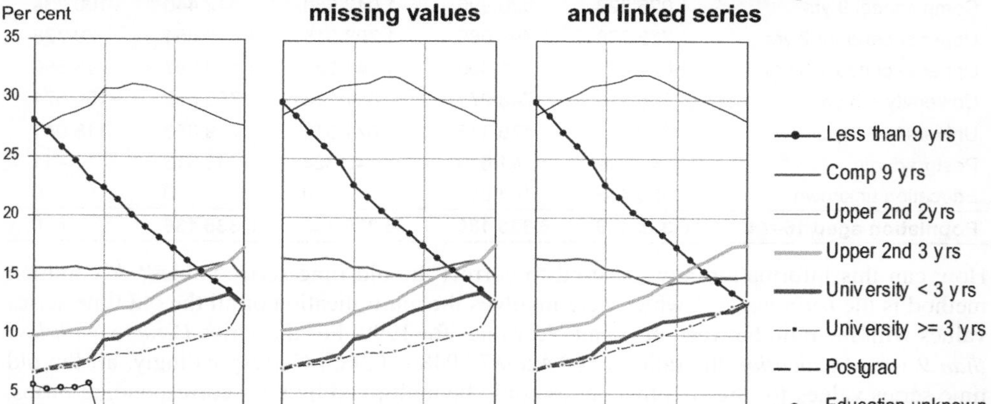

8.3.2 Linking of time series at micro level with weights

It is also possible to use methods with the calibration of weights in this situation. Calibration methodology is used in Section 7.5, where we show how to correct for item nonresponse. The work with correcting time series level shifts can be done in a similar way. Continuing with the example of the Education Register, the work can be carried out in the following stages:

– We assume that 1999 is the first year after the changes, and that 100 important time series exist according to both the old and new levels for 1999 and earlier years.
 100 series have therefore been linked for the period 1985–1999, in the same way as we linked the seven series above. The 100 linked values for every year are completely consistent with one another, which means that all the totals match, e.g. the number of men + the number of women = total number of persons with a particular characteristic.

– It is then possible to calculate the adjusted weight w_i for the objects in the register for every year so that new time series values can be calculated that are comparable with the new time series values for 1999 and onwards. The 100 recalculated values for a particular year are used as the vector t_x in the calibration process. For *every year*, adjusted weights w_i are calculated as described in Section 7.5.

– With the weights w_i all possible time series values for each year can be estimated with consistency and the estimates will then be comparable with values relating to 1999 and onwards.

We have used this method within Statistics Sweden. All monthly Labour Force Surveys for the period 1987–1992 were calibrated to adjust to the changes introduced in 1993. Wallgren (1998) gives a more detailed description of this procedure. Slightly over 100 linked series were used as the calibration conditions and, after a comprehensive check of how the method generates the other linked series, it was discovered that, with appropriate calibration equations, the other series can also be linked in an acceptable manner.

We illustrate below how such calibration works by continuing the example of the Education Register, but only using the seven linked series from the previous section. This provides seven calibration conditions. The table below is taken from Chart 8.20 above, where linking at macro level is discussed.

Chart 8.23 Estimated level shifts

Educational level	1999 new: t_x	1999 old: $\sum d_i g_i x_i$	Level shift: $t_x - \sum d_i g_i x_i$
Less than 9 yrs	862 428	877 948	−15 520
Compulsory school 9 yrs	942 446	1 042 680	−100 234
Upper secondary 2 yrs	1 828 054	1 798 316	29 738
Upper secondary 3 yrs	1 109 997	1 041 337	68 660
University < 3 yrs	760 539	861 213	−100 674
University ≥ 3 yrs	789 359	671 318	118 041
Postgraduate	42 313	42 324	−11

Chart 8.24 below shows how the adjusted weights are calculated using calibration methodology. Columns $x_{1i} - x_{7i}$ show x_i' vectors, one vector for each row. For $i=1$, such as for $PIN1$, $x_1' = (1\ 0\ 0\ 0\ 0\ 0\ 0)$.

The original weights $d_i = 1$ have first been calibrated to adjust for item nonresponse. After this first calibration we have the weights $d_i g_i$ and these weights are now calibrated to adjust for level shifts. The last column in the register shows the adjusted weight w_k, calculated as follows:

1. $T = \Sigma d_i g_i x_i x'_i$ and T^{-1} are calculated, it is the weights $d_i g_i$ that are to be corrected during calibration, where $i = 1, 2, 3, \ldots, 6\ 335\ 136$

   T is the matrix with squared and product totals, here a 7×7 matrix. Because $x_{1i} - x_{7i}$ are either 1 or 0, the matrix is simple with uncorrected numbers for each educational level diagonally and zeros for the rest:

$$T = \begin{bmatrix} 877\ 948 & 0 & 0 & 0 & 0 & 0 & 0 \\ 0 & 1\ 042\ 680 & 0 & 0 & 0 & 0 & 0 \\ 0 & 0 & 1\ 798\ 316 & 0 & 0 & 0 & 0 \\ 0 & 0 & 0 & 1\ 041\ 337 & 0 & 0 & 0 \\ 0 & 0 & 0 & 0 & 861\ 213 & 0 & 0 \\ 0 & 0 & 0 & 0 & 0 & 671\ 318 & 0 \\ 0 & 0 & 0 & 0 & 0 & 0 & 42\ 324 \end{bmatrix}$$

2. The vector λ is calculated: $\lambda = T^{-1}(t_x - \Sigma d_i g_i x_i)$

   The vector t_x constitutes the seven expressions for the numbers that have been corrected for time series level shifts, and the vector $\Sigma d_i g_i x_i$ is the corresponding uncorrected numbers. These numbers are taken from Chart 8.23 above.

3. The adjusted weights become: $w_i = d_i g_i (1 + x'_i \lambda)$

   The adjusted weights give the seven series we have used as calibrations conditions and are used to calculate the other linked time series values.

Chart 8.24 Correction for level shifts in the Education Register 1999

Person	Educational level	x_{1i}	x_{2i}	x_{3i}	x_{4i}	x_{5i}	x_{6i}	x_{7i}	$d_i g_i$	w_i	t_x	$\Sigma d_i g_i x_i$	Difference
PIN1	Less than 9 yrs	1	0	0	0	0	0	0	1.01693	0.99895	862428	877948	−15520
PIN2	Comp school 9 yrs	0	1	0	0	0	0	0	1.01693	0.91917	942446	1042680	−100234
PIN3	Upper 2nd 2 yrs	0	0	1	0	0	0	0	1.01693	1.03375	1828054	1798316	29738
PIN4	Upper 2nd 3 yrs	0	0	0	1	0	0	0	1.01693	1.08398	1109997	1041337	68660
PIN5	University < 3yrs	0	0	0	0	1	0	0	1.01693	0.89805	760539	861213	−100674
PIN6	University ≥ 3 yrs	0	0	0	0	0	1	0	1.01693	1.19574	789359	671318	118041
PIN7	Post-graduate	0	0	0	0	0	0	1	1.01693	1.01667	42313	42324	−11
PIN8	Missing value	0	0	0	0	0	0	0	0	0			
...	...	...	...	...	...	...	...	...	...	...			
PIN6335136	University < 3 yrs	0	0	0	0	1	0	0	1.01693	0.89805			
Total:									6335136	6335136			

The above differences correspond to the corrections for 1999 in Chart 8.21 and 8.23

Correspondingly, calibrated weights for the remaining years' registers are calculated. We show below the calculations for 1998. Adjustment for item nonresponse is carried out with the value 1.01830 for the weight $d_i g_i$.

The seven expressions t_x are the linked values, and $\Sigma d_i g_i x_i$ are the time series values for 1998 that are only corrected for item nonresponse.

Chart 8.25 Correction for level shifts in the Education Register 1998

Person	Educational level	x_{1i}	x_{2i}	x_{3i}	x_{4i}	x_{5i}	x_{6i}	x_{7i}	$d_i\,g_i$	w_i		t_x	$\sum d_i g_i x_i$	Difference
PIN1	Less than 9 yrs	1	0	0	0	0	0	0	1.01830	1.00299		915075	929043	−13968
PIN2	Comp school 9 yrs	0	1	0	0	0	0	0	1.01830	0.92909		939556	1029766	−90210
PIN3	Upper 2nd 2 yrs	0	0	1	0	0	0	0	1.01830	1.03318		1858639	1831874	26764
PIN4	Upper 2nd 3 yrs	0	0	0	1	0	0	0	1.01830	1.08046		1074101	1012306	61794
PIN5	University < 3yrs	0	0	0	0	1	0	0	1.01830	0.90862		750649	841256	−90607
PIN6	University ≥ 3 yrs	0	0	0	0	0	1	0	1.01830	1.18645		749600	643364	106237
PIN7	Post-graduate	0	0	0	0	0	0	1	1.01830	1.01804		39440	39450	−10
PIN8	Missing value	0	0	0	0	0	0	0	0	0				
...	...	...	...	...	...	...	...	...	...	...				
PIN6335136	University < 3 yrs	0	0	0	0	1	0	0	1.01830	0.90862				
Total:									6327060	6327060				

The above differences correspond to the corrections for 1998 in Chart 8.21

An alternative method to correct for level shifts in times series is discussed in Chapter 9.

CHAPTER 9

Estimation with Combination Objects

In this chapter the concept of a *combination object* is introduced. Estimation methods using combination objects are used to reduce *aggregation errors* and correct for level shifts in time series. Aggregation errors may occur when data from different registers with different object types are integrated. In Section 3.3.2, different types of variables, such as single-valued and multi-valued variables, are discussed. A *single-valued variable* can only accept *one* value for each object. A *multi-valued variable* can accept *several* values for certain object. Multi-valued variables are used today in a way that gives rise to aggregation errors. Special estimation methods using weights are introduced in this chapter that can be used for multi-valued variables to reduce these aggregation errors.

9.1 AGGREGATION ERRORS

In Section 3.3.4 aggregation and adjoining are discussed as methods used to create derived variables. To aggregate a qualitative variable is an operation that can give rise to errors and inconsistencies between different registers. In the chart below the total number of employees is *three* in Register 1 but *five* in Register 3. Wage sums by Industry in Register 1 differ from wage sums by Industry in the other registers.

Chart 9.1 Number of employed and wage sums in different registers

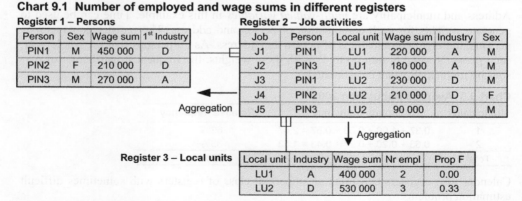

Register 1 – Persons

Person	Sex	Wage sum	1st Industry
PIN1	M	450 000	D
PIN2	F	210 000	D
PIN3	M	270 000	A

Register 2 – Job activities

Job	Person	Local unit	Wage sum	Industry	Sex
J1	PIN1	LU1	220 000	A	M
J2	PIN3	LU1	180 000	A	M
J3	PIN1	LU2	230 000	D	M
J4	PIN2	LU2	210 000	D	F
J5	PIN3	LU2	90 000	D	M

Aggregation

Register 3 – Local units

Local unit	Industry	Wage sum	Nr empl	Prop F
LU1	A	400 000	2	0.00
LU2	D	530 000	3	0.33

The inconsistencies in Chart 9.1 are examples of a general problem that arise when data from different registers are integrated. Even when all variables and identities in the registers

Register-based Statistics – Administrative Data for Statistical Purposes   A. Wallgren and B. Wallgren
© 2007 John Wiley & Sons, Ltd

are entirely correct, these errors will occur. They are created during the integration process. Such errors we call *integration errors*, and the errors discussed in this chapter constitute one kind of integration errors that we call *aggregation errors*.

When we aggregate variable values from *many objects to one object* this kind of error will occur if the variable is qualitative. The same problem arises if one object can occur several times in a register, but with different characteristics, e.g. when an object has changed during the time period that is the reference period of the register. The problem with aggregation errors can arise from three reasons, discussed below:

– objects occur several times in a register,

– many to one relations, and

– multi-valued variables.

Objects occurring several times – calendar year registers
In a calendar year register, objects that change during the year occur several times. Persons move or change civil status, households are changed; enterprises split or merge or change their branch of industry. Some objects can change many times during the year and all these changes give rise to multi-valued variables. The example in Chart 7.8 is continued in the chart below, where persons move during 2005.

Chart 9.2 Calendar year register for the population of persons during 2005

Person	Address	Municipality	From date yyyymmdd	To date yyyymmdd	Weight = Time at the address, years
PIN1	Address 1	1	20050101	20050517	136/365 = 0.37
PIN1	Address 2	2	20050518	20051231	229/365 = 0.63
PIN2	Address 3	1	20050101	20051231	365/365 = 1.00
PIN3	Address 4	2	20050101	20050314	73/365 = 0.20
PIN3	Address 5	1	20050315	20050925	194/365 = 0.53
PIN3	Address 6	2	20050926	20051231	365/365 = 0.27
PIN4	Address 7	2	20050101	20050605	156/365 = 0.43
PIN4	Address 8	1	20050606	20051231	209/365 = 0.57

Address and municipality are *multi-valued variables* in this example. Time can be used to generate weights for each combination of person and address. The register contains *four* persons, but *eight* combinations of person and address. As all persons live the whole year, each person's weights sum up to 1. With these weights, the frequency distribution of persons by municipality can be estimated.

Chart 9.3 Average population 2005

Municipality	Absolute frequency	Relative frequency
1	0.37 + 1.00 + 0.53 + 0.57 = 2.47	62%
2	0.63 + 0.20 + 0.27 + 0.43 = 1.53	38%
Total	4.00	100%

Calendar year registers constitute an important class of registers with sometimes difficult estimation problems.

Many to one relations
In the example above, there is *one* register where some objects appear many times, and this gives rise to multi-valued variables. In other cases *two* registers with different object types are matched, and when there are many-to-one relations, multi-valued variables can be

created when these registers are integrated and qualitative variables are aggregated. If a person has many jobs, how should information about these jobs be aggregated into information about the person? The example in Chart 9.1 can be used as an illustration. Register 2 is now sorted by PIN.

Chart 9.4 Number of employed and wage sums in different registers

Register 1 – Persons

Person	Sex	Wage sum	1st Industry
PIN1	M	450 000	D
PIN2	F	210 000	D
PIN3	M	270 000	A

Aggregation

Register 2 – Job activities

Job	Person	Local unit	Wage sum	Industry	Sex
J1	PIN1	LU1	220 000	A	M
J3	PIN1	LU2	230 000	D	M
J4	PIN2	LU2	210 000	D	F
J2	PIN3	LU1	180 000	A	M
J5	PIN3	LU2	90 000	D	M

In register 2, Industry is a single-valued variable describing a characteristic of the object type job or activity. PIN1 and 3 have two jobs, both persons work at a local unit within industry A, and also at a local unit within industry D. The traditional way to create a variable Industry for persons is to use information about only *one* job for each person – the most important job. In this situation, when most persons have one job but some have more than one, but a limited number of jobs, a better solution is to define the variables *local unit* and *Industry for persons*, as multi-valued variables. In register 2 in Chart 9.4 above, these variables are single-valued variables for *jobs*, but in Chart 9.5 below, these variables have been transformed into multi-valued variables for *persons*. The wage sums for each person are used to create weights, where the weights for each person sum up to 1.

Chart 9.5 Number of employed and wage sums in different registers

Person	Sex	Local unit	Wage sum	Industry	Weight
PIN1	M	LU1	220 000	A	22/45 = 0.49
PIN1	M	LU2	230 000	D	23/45 = 0.51
PIN2	F	LU2	210 000	D	21/21 = 1.00
PIN3	M	LU1	180 000	A	18/27 = 0.67
PIN3	M	LU2	90 000	D	9/27 = 0.33

The data matrix in Chart 9.5 can be used to estimate tables with persons by industry.

Chart 9.6 Employed by Industry

Industry	Number of employed
A	0.49 + 0.67 = 1.16
D	0.51 + 1.00 + 0.33 = 1.84
Total	3.00

Both examples in this section show that multi-valued variables can arise for different reasons. In the next section, multi-valued variables in register systems are discussed and estimation methods are proposed, which leads to consistent estimates. The inconsistencies in Chart 9.1 become unnecessary if these methods are used.

9.2   ESTIMATION METHODS FOR MULTI-VALUED VARIABLES

The variable *Highest Education* is created in the Education Register, which is a register on persons. This variable is a multi-valued variable, as some persons have two or more degrees on the same level. The variable *Industry* is created in the Business Register, and is also a multi-valued variable of great importance. Both these are examples of variables which are multi-valued at the original source. As mentioned in the previous section, multi-valued variables are also created in the register system.

Multi-valued variables are difficult to deal with, but they are both common and important within the register system. Multi-valued variables in statistical registers are also used in censuses and sample surveys, which means that the problem will also affect these types of survey.

These problems are usually 'solved' in a drastic way – the multi-valued variable is transformed into a single-valued variable by only using 'the most important value' for every object. If, for example, the distribution of persons by different occupations is to be described, the occupations that are common as secondary activities will be underestimated. A portion of the occupational information is discarded, and estimates will then have quality problems of an unknown magnitude due to aggregation errors.

We begin with a simple example to show the fundamental principles for how we believe that multi-valued variables should be treated. We then look at more complicated situations that occur when the principles are used in practice.

9.2.1   Occupation in the Activity and Occupation Registers

The estimation problem in this section can be defined as follows: how should the frequency distribution of different occupations be estimated?

> *Principle 1:*
> What calculations should be done? The estimation problem should always be specified before the calculations begin. This is illustrated below where three ways of defining and solving the estimation problem are compared.

The following data matrix in chart 9.7 shows occupation and occupational code (ISCO) for six persons, of whom two have more than one occupation. The object in the matrix is *job*, which is a *relational object* that is identified by personal identification number and the legal unit identity for each enterprise. The variable *Extent*, the extent of the work, is given as a percentage of full-time work. This variable is taken from the salary register. Suppose that the data matrix contains all occupational activities in a small region, how should we estimate the distribution of *persons* into the different occupations? This is our first estimation problem.

Chart 9.7 Job Register with occupational data

Job id	Person	Legal unit	Occupation	ISCO	Extent
J1	PIN1	LeU1	Statistician	2211	100
J2	PIN1	LeU2	Farmer	6111	15
J3	PIN1	LeU3	Politician	1110	10
J4	PIN2	LeU4	Hospital orderly	5132	30
J5	PIN2	LeU5	Cleaner	9122	20
J6	PIN3	LeU6	Shop assistant	5221	10
J7	PIN4	LeU6	Shop assistant	5221	50
J8	PIN5	LeU6	Shop assistant	5221	20
J9	PIN6	LeU6	Shop assistant	5221	100
Total					

The traditional approach is that each person has only *one* occupation – the *principal occupation*. This means that information for those with several occupations is discarded; only the occupation with the largest extent of work is included.

We then get a new data matrix (Chart 9.8), in which the object is *person* and where the distribution by occupation is acquired by summing the number of persons in each occupation.

Chart 9.8 Traditional register on persons with occupational information

Person	Legal unit	Principal occupation	ISCO	Extent	Weight alt 1
PIN1	LeU1	Statistician	2211	100	1
PIN2	LeU4	Hospital orderly	5132	30	1
PIN3	LeU6	Shop assistant	5221	10	1
PIN4	LeU6	Shop assistant	5221	50	1
PIN5	LeU6	Shop assistant	5221	20	1
PIN6	LeU6	Shop assistant	5221	100	1
Total					6

Chart 9.9 below contains the estimated occupational distribution. According to the tradition within statistics on persons, every person has the same weighting regardless of whether they work 100% or 10% of a full-time job.

Chart 9.9 Employed persons by occupation, traditional alternative 1

Main occupation	ISCO	Number	Per cent
Statistician	2211	1	16.7
Hospital orderly	5132	1	16.7
Shop assistant	5221	4	66.7
Total		6	100.0

This example shows that occupations that are common secondary occupations, such as the occupations *politician* and *farmer* that are often undertaken alongside the principal occupation, are underestimated. Estimates for multi-valued variables can instead be made in a way that avoids discarding any information. This is possible if estimates are based on a data matrix with 'combination objects'.

> *Principle 2:*
> The basic principle is to create a data matrix so that *every combination of object and value of the multi-valued variable corresponds to one object* in the new data matrix. Objects, or rows, in such data matrices are called *combination objects*.

The data matrix in the chart below has been formed in this way; the six persons in the register on persons in Chart 9.8 give rise to nine combination objects.

Chart 9.10 Register on persons with occupational data

Combination object	Person	Occupation	Extent	Weight alternative 1	Weight alternative 2
1	PIN1	Statistician	100	1	0.80
2	PIN1	Farmer	15	0	0.12
3	PIN1	Politician	10	0	0.08
4	PIN2	Hospital orderly	30	1	0.60
5	PIN2	Cleaner	20	0	0.40
6	PIN3	Shop assistant	10	1	1.00
7	PIN4	Shop assistant	50	1	1.00
8	PIN5	Shop assistant	20	1	1.00
9	PIN6	Shop assistant	100	1	1.00
Total				6	6.00

The rows in the matrix consist of all combinations of *person • occupation*. For example, person *PIN1*, who has three occupations, appears in three rows in the matrix. The weights according to alternative 2 have been calculated with the variable *Extent* so that 0.80 = 100/(100 + 15 + 10) etc. The weights for each person sum up to 1 in both alternatives 1 and 2. All weights in both alternatives sum up to 6, i.e. the total number of persons. Chart 9.10 above illustrates several general principles:

> *Principle 3:*
> The sum of the weights for *one person* (the object type that the estimation problem refers to and that was the starting point when forming the combination objects) should always be 1.

> *Principle 4:*
> It follows from principle 3 that the sum of all the weights is the same as the total number of objects (the object type that the estimation problem refers to).

In the following chart, the frequency distribution of persons by occupation is calculated with weights according to alternatives 1 and 2 in Chart 9.10.

Chart 9.11 Employed persons by occupation according to two alternatives

Occupation	ISCO	Alternative 1		Alternative 2	
		Nr.	Per cent	Nr.	Per cent
Politician	1110	0.00	0.0	0.08	1.3
Statistician	2211	1.00	16.7	0.80	13.3
Hospital orderly	5132	1.00	16.7	0.60	10.0
Shop assistant	5221	4.00	66.7	4.00	66.7
Farmer	6111	0.00	0.0	0.12	2.0
Cleaner	9122	0.00	0.0	0.40	6.7
Total		**6.00**	**100.0**	**6.00**	**100.0**

The number of employed persons (Nr.) by occupation is estimated by summing up the weights for each occupation. Weight alt 1 in Chart 9.10 is summed up in alternative 1 and Weight alt 2 is summed up in alternative 2.

Because some of the occupations have the weight 0 in alternative 1, corresponding to the traditional method of calculation, aggregation errors occur; estimates according to alternative 1 are distorted, in that the frequency of certain occupations is overestimated while the frequency of others is underestimated. However, estimates according to alternative 2 utilise all the information on the occupations in the multi-valued variable.

The weights in alternative 2 utilise the variable *Extent*. This variable is found in the Swedish Salary Register; for those positions that are not included there, weights must be formed from other information. The Statement of Earnings Register contains *annual gross wages* for all jobs, and can therefore always be used. Weights calculated from annual gross wages are somewhat different to weights calculated from *Extent*. When working to create good estimates, it is necessary to choose between different *weight-generating variables* and it is important to choose a variable that is both relevant and functional.

The weights according to alternatives 1 and 2 are both based on the variable *Extent*, but could also be based on other variables. The weights actually used can differ more or less from the ideal weights; for certain persons, the weights for one occupation can be too large, while for others, the weights for the same occupation can be too small. The errors can partly be balanced out when forming the distribution of occupations overall. The relevant quality measurement could be a measurement of how close the estimated distribution is to the distribution that would be calculated with ideal weights.

> *Principle 5:*
> It is better to use good weights rather than bad, even if the good weights are not entirely perfect.

In alternatives 1 and 2, the estimation problem is to describe the distribution of *persons* by occupation. A third alternative, *alternative 3,* is to distribute the *extent of work* by occupation. Extent or volume of work could be described by the amount of occupational activity recalculated to full-time employed persons. This method of calculating is common in economic statistics where it is usual to measure volumes instead of persons.

Person *PIN1* has three occupations, one full-time and the two other corresponding to 15% and 10% of a full-time employed position. The matrix with the six persons represents 3.55 *full-time employed positions.* The object in the matrix in Chart 9.12 is *job*, and the variable occupation is a single-valued variable – every job corresponds to only one occupation. The distribution of full-time employed positions by occupation is given by summing up the variable *Weight alt 3 (= Extent /100 in Chart 9.10)* for the different occupations.

Chart 9.12 Register on persons with occupational data

Person	Occupation	Weight alt 1	Weight alt 2	Weight alt 3
PIN1	Statistician	1	0.80	1.00
PIN1	Farmer	0	0.12	0.15
PIN1	Politician	0	0.08	0.10
PIN2	Hospital orderly	1	0.60	0.30
PIN2	Cleaner	0	0.40	0.20
PIN3	Shop assistant	1	1.00	0.10
PIN4	Shop assistant	1	1.00	0.50
PIN5	Shop assistant	1	1.00	0.20
PIN6	Shop assistant	1	1.00	1.00
Total		**6**	**6.00**	**3.55**

Chart 9.13 Persons and full-time employed by occupation, three alternatives

Occupation	ISCO	Alternative 1		Alternative 2		Alternative 3	
		Nr.	Per cent	Nr.	Per cent	Nr.	Per cent
Politician	1110	0.00	0.0	0.08	1.3	0.10	2.8
Statistician	2211	1.00	16.7	0.80	13.3	1.00	28.2
Hospital orderly	5132	1.00	16.7	0.60	10.0	0.30	8.5
Shop assistant	5221	4.00	66.7	4.00	66.7	1.80	50.7
Farmer	6111	0.00	0.0	0.12	2.0	0.15	4.2
Cleaner	9122	0.00	0.0	0.40	6.7	0.20	5.6
Total		**6.00**	**100.0**	**6.00**	**100.0**	**3.55**	**100.0**

Alternatives 1 and 2 relate to the same estimation problem, *persons* distributed by occupation, but are based on different estimation methods that use different weights.

Alternative 3 relates to another estimation problem, *extent of work* distributed by occupation. The focus is here on the volume of work, not on persons.

9.2.2 Industrial classification in the Business Register

Industrial classification is another important multi-valued variable. It is created in the Business Register, and is used by many registers within the register system. Here, it is also common to select the 'most important industrial classification' and discard information on other industrial classifications for the local units or enterprise units the statistics refer to. This leads to aggregation errors in economic statistics.

The Business Register at Statistics Sweden contains information on all branches of industry in which an enterprise is involved. There are also details of the proportion of business carried out within each industry. Industrial classification code and the share within each industry are of good quality when relating to manufacturing enterprises. The method of choosing the most important industry can cause problems when reporting industry statistics and can also cause time series problems. For example, if 51% of the activities in a large enterprise in year 1 fall within a particular industrial classification, but only 49% fall within the same industrial classification in year 2, this small change can give significant level shifts in many time series, in which, for example, all employees working at this large enterprise seemingly change industrial classification from year 1 to year 2.

Within regional statistics, these problems can be even more serious, as one local unit can be dominating, which means that a change in industrial classification will cause time series level shifts in regional series.

Slight changes are even more problematic as they are harder to detect, and in many cases will be misinterpreted as real changes in the economy. The methodology presented in the previous section, relating to occupation, makes it possible to avoid these quality problems.

In the Chart 9.14 below, industrial classification and number of employees are shown for three local units. *The estimation problem relates to estimating the number of employees by industrial classification.* The information used by the traditional method is shown in the shaded table cells; the information in the unshaded cells is available but is not used.

Chart 9.14a Business Register year 1: Data matrix for local units

Local unit	Industry 1	%	Industry 2	%	Industry 3	%	Nr. of employees
LU1	DJ	100					218
LU2	DH	51	DJ	49			293
LU3	DJ	40	DH	30	DK	30	156

Chart 9.14b Business Register year 2: Data matrix for local units

Local unit	Industry 1	%	Industry 2	%	Industry 3	%	Nr. of employees
LU1	DJ	100					221
LU2	DJ	52	DH	48			314
LU3	DJ	36	DH	34	DK	30	143

Chart 9.14c Number of employees by industry, traditional estimates

Industry	Year 1	Year 2
DH	293	0
DJ	374	678
DK	0	0
Total	**667**	**678**

The number of employees is sorted by principal industry, which is the most common way of presenting time series based on industrial classification from the Business Register.

This leads to abrupt changes in the series here.

In Section 9.2.7 there is an example illustrating the time series problems, which can be caused by the traditional estimation methods. That example is based on actual data and the time series disturbance is serious.

In Charts 9.14a and 9.14b, the percentages show the share of each industry, which is a measure of the size of every industry at every local unit. The size measurement can be based on turnover, number of employees or something else.

We first assume that the percentage charts are based on number of employees and then show how the weights, based on a specific size measure, are transformed into weights based on another size measure.

According to *principle* 2 in the previous section, a new data matrix (see Chart 9.15 below) is created containing combination objects, so that every combination of objects and values for the multi-valued variable corresponds to one row in the new data matrix.

In this data matrix, every row is a combination of local unit and industrial classification. Instead of a matrix with *three* rows referring to three local units, we get a new matrix with *six* rows referring to all combinations of *Local unit • Industry* for each year. For every industry, we can then estimate the number of employees with formula (4):

$$\hat{Y} = \sum_{i=1}^{R} w_i y_i \qquad \text{In Chart 9.15 below, } w_i y_i \text{ has been calculated for every row} \qquad (4)$$

In the previous section with occupation, the variable $y_i = 1$ for all combination objects. Formula (4) then indicates that the weights w_i are summed up for every table cell in Chart 9.13.

Chart 9.15 Data matrix with combination objects: Local unit • Industry

Year 1					Year 2				
Local unit	Industry	Weight, w_i	Nr. empl, yi	$w_i y_i$	Local unit	Industry	Weight, w_i	Nr. empl, yi	$w_i y_i$
LU1	DJ	1.00	218	218	LU1	DJ	1.00	221	221
LU2	DH	0.51	293	149.43	LU2	DH	0.48	314	150.72
LU2	DJ	0.49	293	143.57	LU2	DJ	0.52	314	163.28
LU3	DJ	0.40	156	62.4	LU3	DJ	0.36	143	51.48
LU3	DH	0.30	156	46.8	LU3	DH	0.34	143	48.62
LU3	DK	0.30	156	46.8	LU3	DK	0.30	143	42.9
Total		**3.00**		**667**	**Total**		**3.00**		**678**

The weights w_i sum up to 3, as we are still talking about three local units. The sums of the products $w_i y_i$ will be the same total as before:

The total number of employees is a given total that should not be changed when introducing weights. The weights will only affect how the employees are distributed between different industries.

This is an example of a principle that is generally applicable.

Chart 9.16 Number of employees by industry, estimated with combination objects

Industry	Year 1	Year 2
DH	196.23	199.34
DJ	423.97	435.76
DK	46.80	42.90
Total	**667.00**	**678.00**

The time series in Chart 9.16 have been calculated with the weights w_i.

The series here have a higher quality than those in Chart 9.14c, with relevant changes and no level shifts.

9.2.3   Transformation of weights

The weights for the different industrial classifications in the Swedish Business Register are primarily based on turnover. They should be transformed when calculating estimates for other variables.

Example: When estimating number of employees, the weights based on turnover should be transformed into weights that are suitable for number of employees. This transformation is based on a model that describes the relation between turnover and number of employees, and the model is based on appropriate statistics that describe employment and turnover for industry-specific local units. Chart 9.17 below shows this transformation for local unit *LU3*. The weight based on turnover is multiplied by the number of employees per turnover. These values must then be divided by a constant so that the total is 1 for every local unit.

Chart 9.17 Transformation of weights

Register of local units			Aggregated data		Register of local units		
Year 1		Weights	Models for different industries		Transformed weights based on model		
Local unit	Industry	based on turnover	$\dfrac{\text{Employees}}{\text{Turnover SEK m}}$		adapted for estimation of number of employees		
LU3	DJ	0.4	DJ	0.5	LU3	DJ	$\dfrac{0.4 \cdot 0.5}{(0.4 \cdot 0.5 + 0.3 \cdot 0.6 + 0.3 \cdot 0.7)} = \mathbf{0.34}$
LU3	DH	0.3	DH	0.6	LU3	DH	$\dfrac{0.3 \cdot 0.6}{(0.4 \cdot 0.5 + 0.3 \cdot 0.6 + 0.3 \cdot 0.7)} = \mathbf{0.30}$
LU3	DK	0.3	DK	0.7	LU3	DK	$\dfrac{0.3 \cdot 0.7}{(0.4 \cdot 0.5 + 0.3 \cdot 0.6 + 0.3 \cdot 0.7)} = \mathbf{0.36}$

According to principle 3, the sum of the weights for *one local unit* (the object type the estimation problem refers to) should always be 1.

The transformed weight for the share of the local unit belonging to the capital-intensive steel industry DJ is lower than the original weight, which was based on turnover (0.34 compared to 0.4). The example also shows that, when using turnover, the most important industrial classification is DJ, but when using number of employees, it is presumably DK.

This again shows that the principle of only using the 'most important' value of a multi-valued variable can cause problems. Furthermore, when registers contain multi-valued variables, weights should be calculated that are adapted for different estimation problems.

9.2.4   Importing many multi-valued variables

Many variables from several registers are imported into the Employment Register. The population is created using the Population Register on December 31, from where single-valued variables such as age and sex are also taken. The multi-valued variables education and occupation are taken from the Education Register and the Occupation Register, respectively.

The Activity Register contains the object *job*, which is a relation between person and local unit. The industrial classification of the local unit is imported from the Business Register into the Activity Register. Finally, local unit identity, together with industrial classification is exported from the Activity Register into the Employment Register. Both local unit identity, and industrial classification are multi-valued variables in the Employment Register.

A derived variable is created in the Employment Register, showing whether a person was gainfully employed during November (called *EmpNov* below). The method by which this variable is created is described in Section 6.2.3.

With traditional estimation methodology, only data referring to the 'most important' variable value are used for all these multi-valued variables (shaded in the data matrices in Chart 9.18 below).

We first show all the involved registers with the data linked to a specific person *PIN10* and how the traditional estimate is made. After that we show how all the information can be used with combination objects for the multi-valued variables.

1. Traditional methodology – only the most important value is used

Chart 9.18a Population Register

Person	Sex	Age
PIN10	F	32

Chart 9.18b Education Register

Person	Educ 1	Points 1	Educ 2	Points 2
PIN10	Ed1	180	Ed2	120

PIN10 has two degrees on the same level in different fields; Educ 2 is the most recent.

Chart 9.18c Activity Register, with extent of job in November

Person	Local unit	Extent
PIN10	LU11	80%
PIN10	LU12	20%

Traditionally, only the local unit of the principal activity is used.

Chart 9.18d Occupation Register

Person	Local unit	Occup.
PIN10	LU11	Oc1
PIN10	LU12	Oc2

Traditionally, only the occupation of the principal activity is used.

Chart 9.18e Business Register

Local unit	Industry	Weight 1	Industry	Weight 2
LU11	DH	70%	DJ	30%
LU12	DK	100%		

The Activity and Business Registers are matched using the local unit identity as the linkage variable. The largest *Industry* is imported to the Activity Register. The Activity Register is then matched with the Occupation Register using the personal identification number and local unit identity as linkage variables. *Occupation* is imported to the Activity Register.

Chart 9.18f Activity Register, industry and occupation are imported

Person	Local unit	Industry	Occup.
PIN10	LU11	DH	Oc1
PIN10	LU12	DK	Oc2

Traditionally, only the local unit and occupation of the principal activity are used.

The Employment Register is created in three stages:

– A new register is created with persons aged 16 and over, with the variables *sex* and *age* from the Population Register.

– The most important variable values for the multi-valued variables education, *occupation*, *local unit* and *industrial classification* are imported from the different registers.

– The variable *EmpNov*, relating to gainfully employed persons in November, is created.

The part of the completed Employment Register relating to person *PIN10* looks as shown in Chart 9.18g.

Chart 9.18g Employment Register, data for person PIN10

Person	Sex	Age	Education	Occupation	Local unit	Industry	EmpNov
PIN10	F	32	Ed2	Oc1	LU11	DH	Yes

2. Methodology with combination objects, when all information is used

Chart 9.19a Population Register

Person	Sex	Age
PIN10	F	32

Chart 9.19b Education Register

Person	Education	w_{Edu}
PIN10	Ed1	0.6
PIN10	Ed2	0.4

Weights for education are created using the length of the educational programme expressed as education 'points'.

Chart 9.19c Activity Register, with extent of job in November

Person	Local unit	w_{LU}
PIN10	LU11	0.8
PIN10	LU12	0.2

For the object *Person*, weights for the multi-valued variable *Local unit* are created using the variable *Extent*

Chart 9.19d Occupation Register

Person	Local unit	Occup.
PIN10	LU11	Oc1
PIN10	LU12	Oc2

Occupation is linked to the relation between *Person* and *Local unit*, the weight for *Occupation* is the same as that for *Local unit*.

Chart 9.19e Business Register

Local unit	Industry	w_{Ind}
LU11	DH	0.7
LU11	DJ	0.3
LU12	DK	1.0

Using information in the Business Register, a register is created with the combination object *Local unit · Industry* and the weights for different industries.

An Employment Register is formed with combination objects that are combinations of the relevant object type *Person* and *all* values of *all* multi-valued variables, according to the principle mentioned above. (*Principle 2:* Create a data matrix with combination objects so that *every combination of objects and values of the multi-valued variable corresponds to one row* in the new data matrix.)

As a result of the matches, a data matrix is created. The part of the data matrix referring to person *PIN10* is shown in Chart 9.19f.

Chart 9.19f Combination objects: Person · Education · Local unit · Industry

Person	Sex	Age	Educ.	Local unit	Occup	Industry	EmpNov	w_{Edu}	w_{LU}	w_{Ind}	$w_{CombObj}$
PIN10	F	32	Ed1	LU11	Oc1	DH	Yes	0.6	0.8	0.7	0.336
PIN10	F	32	Ed1	LU11	Oc1	DJ	Yes	0.6	0.8	0.3	0.144
PIN10	F	32	Ed1	LU12	Oc2	DK	Yes	0.6	0.2	1.0	0.120
PIN10	F	32	Ed2	LU11	Oc1	DH	Yes	0.4	0.8	0.7	0.224
PIN10	F	32	Ed2	LU11	Oc1	DJ	Yes	0.4	0.8	0.3	0.096
PIN10	F	32	Ed2	LU12	Oc2	DK	Yes	0.4	0.2	1.0	0.080
Total											1.000

The calculations should refer to the object type *Person*. The variables *Education, Local unit* and *Occupation* are multi-valued variables for the object type *Person*. *Industry* is a multi-valued variable for the object type *Local unit*. Every combination of *Person* and *Local unit* corresponds to only one *Occupation*, which is why no additional weight is needed for the variable *Occupation. Occupation* is here a single-valued variable for every combination of *Person* and *Local unit.*

> **Principle 6:**
> When there are many multi-valued variables in the same data matrix, all the weights for the multi-valued variables are multiplied to obtain the weights that are to be used for estimation.

Chart 9.19f shows person *PIN10* divided up into six combination objects or rows. The weights $w_{CombObj}$ for these six combination objects should sum up to 1 and are formed by multiplication, for example, $0.6 \cdot 0.8 \cdot 0.7 = 0.336$

3. Traditional estimation compared with estimation with combination objects

Using the variables age, sex, education, occupation and industry, frequency tables can be formed for the variable *EmpNov* (gainfully employed in November). Starting from the data matrix in Chart 9.19f above, we now show how person *PIN10 contributes to the estimates* of the frequencies in the different table cells.

The tables on the left in Chart 9.20 show the frequencies estimated in the traditional way. Corresponding tables on the right show the frequencies estimated by summing the weights $w_{CombObj}$ in each cell. There are many differences between the estimation methods; with the traditional method, much of the information for the multi-valued variables is discarded which leads to aggregation errors.

Chart 9.20a Number of gainfully employed persons in November, by age and sex

Traditional estimation

Age	F	M	Total
20–49	1	0	1
50–64	0	0	0
65–	0	0	0
Total	1	0	

The estimates for single-valued variables such as sex and age are not affected by the weights that are formed for the multi-valued variables.

Estimation with weights

Age	F	M	Total
20–49	1	0	1
50–64	0	0	0
65–	0	0	0
Total	1	0	1

Chart 9.20b Number of gainfully employed in November by occupation

Traditional estimation

Occup.	Number
Oc1	1
Oc2	0
Total	1

Estimation with weights:
$0.8 =$
$0.336 + 0.144 + 0.224 + 0.096$

Estimation with weights

Occup.	Number
Oc1	0.8
Oc2	0.2
Total	1

Chart 9.20c Number of gainfully employed in November by education

Traditional estimation

Educ.	Number
Ed1	0
Ed2	1
Total	1

Estimation with weights:
$0.6 = 0.336 + 0.144 + 0.120$

Estimation with weights

Educ.	Number
Ed1	0.6
Ed2	0.4
Total	1

Chart 9.20d Gainfully employed in November by industrial classification

Traditional estimation

Industry	Number
DH	1
DJ	0
DK	0
Total	1

Estimation with weights:
$0.56 = 0.336 + 0.224$

Estimation with weights

Industry	Number
DH	0.56
DJ	0.24
DK	0.20
Total	1

Chart 9.20e Gainfully employed in November by occupation and education

Traditional estimation

Educ.	Oc1	Oc2	Total
Ed1	0	0	0
Ed2	1	0	1
Total	1	0	1

Estimation with weights:
0.48 = 0.336 + 0.144

Estimation with weights

Educ.	Oc1	Oc2	Total
Ed1	0.48	0.12	0.60
Ed2	0.32	0.08	0.40
Total	0.80	0.20	1

Chart 9.20f Gainfully employed in November by education and industry

Traditional estimation

Industry	Ed1	Ed2	Total
DH	0	1	1
DJ	0	0	0
DK	0	0	0
Total	0	1	1

Estimation with weights:
0.336 is taken directly from the data matrix in Chart 9.19f

Estimation with weights

Industry	Ed1	Ed2	Total
DH	0.336	0.224	0.560
DJ	0.144	0.096	0.240
DK	0.120	0.080	0.200
Total	0.600	0.400	1

Chart 9.20g Gainfully employed in November by occupation and industry

Traditional estimation

Industry	Oc1	Oc2	Total
DH	1	0	1
DJ	0	0	0
DK	0	0	0
Total	1	0	1

Estimation with weights
0.56 = 0.336 + 0.224

Estimation with weights

Industry	Oc1	Oc2	Total
DH	0.56	0.00	0.56
DJ	0.24	0.00	0.24
DK	0.00	0.20	0.20
Total	0.80	0.20	1

9.2.5 Consistency between different variables

To ensure consistency when using the different multi-valued variables in the system, the weights for these should be included in the register that is responsible for each respective multi-valued variable. Everyone should then use these weights.

Certain registers contain many variables that need to fulfil certain consistency conditions, such as records in a profit and loss statement. Consistency is maintained if the same weights are used for all variables, or if sub-records are recalculated with different weights first, and then the totals and differences are calculated.

9.2.6 Consistency between estimates from different registers

When several registers contain the same multi-valued variable, the estimates from different registers should be equal. First, all register populations must be consistent. In the example below this is the case; the persons are the same both in the Job Register and the register on persons. Furthermore, local units and enterprise units are the same in both the Job Register and the Business Register.

Secondly, all the variables must be consistent. In the example below, the variables *Industry* and *Ext* (extent of work) are consistent in the different registers; the same variable definitions and measurement errors exist in all registers as the variables have been imported from the original register to other registers.

The third condition for consistency is that estimates are made with weights so that all information from the multi-valued variable *Industry* is included in the estimates. If only the most important value of *Industry* is used instead of weights, the estimates will contain errors and *these aggregation errors will differ for the different registers so that the estimates will be inconsistent.* This is shown in the example below.

1. Integration of data from four registers – available data

The example consists of a register system with four registers: a Register on five persons (PIN1–PIN5), a Job Register with six jobs (J1–J6), a Local unit Register with 3 local units (LU11, LU21, LU22) and an Enterprise Register with two enterprise units (EU1, EU2). In Chart 9.21 all available information is shown before any integration of data has been done.

Chart 9.21 Available information in four registers before integration

1. Register on persons
Person
PIN1
PIN2
PIN3
PIN4
PIN5

2. Job Register				
Job	Enterprise	Local unit	Person	Ext_J
J1	EU1	LU11	PIN1	0.3
J2	EU1	LU11	PIN2	1.0
J3	EU2	LU21	PIN3	1.0
J4	EU2	LU21	PIN4	1.0
J5	EU2	LU22	PIN5	1.0
J6	EU2	LU22	PIN1	0.2

Person PIN1 has two jobs, J1 and J6, PIN1 works 50% of a full-time employed position.

The information of the proportions of Industries within each Local Unit will be imported into registers 1 and 2 when combination objects are created and the aggregated weight will be called w_{Ind} in Chart 9.23 below.

3. Local unit Register					
Local unit	Enterprise	Industry 1	%	Industry2	%
LU11	EU1	A	60	B	40
LU21	EU2	C	100		
LU22	EU2	D	100		

4. Enterprise Register						
Enterprise	Local unit 1	Local unit 2	Industry1	%	Industry2	%
EU1	LU11		A	60	B	40
EU1	LU21	LU22	C	62.5	D	37.5

2. Traditional estimates – aggregation errors

Chart 9.22 shows the traditional way of calculating full-time employees by Industry. In all registers, only *largest Industry* is used for each local unit, enterprise unit, job or person. Thus *Industry* is forced to become a single-valued variable – the estimates of the number of full-time employees by Industry will be different due to different aggregation errors.

Chart 9.22 Traditional estimation in a register system after integration

1. Employment Register – persons			
Person	Industry	Ext_P	weight
PIN1	A	0.5	1
PIN2	A	1.0	1
PIN3	C	1.0	1
PIN4	C	1.0	1
PIN5	D	1.0	1
Total		4.5	5

2. Job Register						
Job	Enterprise	Local unit	Person	Ext_J	Industry	weight
J1	EU1	LU11	PIN1	0.3	A	1
J2	EU1	LU11	PIN2	1.0	A	1
J3	EU2	LU21	PIN3	1.0	C	1
J4	EU2	LU21	PIN4	1.0	C	1
J5	EU2	LU22	PIN5	1.0	D	1
J6	EU2	LU22	PIN1	0.2	D	1
Total				4.5		6

In all the registers in Chart 9.22 the variable Industry refers to the *principal* Industry. Every register also has only one row per object (person, job, local unit or enterprise unit).

The variable Ext_J, Extent for Job, in the Job Register is imported into all other registers. Derived variables Ext_P, Extent for Person, Ext_{LU}, Extent for Local Unit and Ext_{EU}, Extent for Enterprise Unit, are created by aggregation.

With these four registers, volume of work by Industry, can be estimated by summing up the variables:

Ext_P, Ext_J, Ext_{LU} and Ext_{EU}

Due to different aggregation errors, the estimated tables with volume of work by Industry will be different if different registers are used.

These different tables are compared in Chart 9.24 below.

3. Local unit Register			
Local unit	Industry	Ext_{LU}	weight
LU11	A	1.3	1
LU21	C	2.0	1
LU22	D	1.2	1
Total		4.5	3

4. Enterprise Register				
Enterprise	Local unit	Industry	Ext_{EU}	weight
EU1	LU11	A	1.3	1
EU2	LU21	C	3.2	1
Total			4.5	2

3. Estimation with combination objects

In all registers, combination objects are formed by *object · Industry* and weights are calculated for the combination objects. In Chart 9.23 below, the variables *Industry* and w_{Ind} have been imported from the Local unit Register to the Job Register. By summing up $Ext_J · w_{Ind}$ in the Job Register, *Extent for persons, Extent for local units* and *Extent for enterprise* units are obtained.

Chart 9.23 Consistent estimates with weights in a register system

1. Employment Register – persons					
Person	Industry	Ext_P w_{Ind} w_{Job}		$w_P =$ $w_{Ind} ·$ w_{Job}	$Ext_P ·$ w_P
PIN1	A	0.50	0.6	0.6	0.36	0.18
PIN1	B	0.50	0.4	0.6	0.24	0.12
PIN1	D	0.50	1	0.4	0.4	0.20
PIN2	A	1.00	0.6	1	0.6	0.60
PIN2	B	1.00	0.4	1	0.4	0.40
PIN3	C	1.00	1	1	1	1.00
PIN4	C	1.00	1	1	1	1.00
PIN5	D	1.00	1	1	1	1.00
Total					5	4.5

2. Job Register							
Job	Enterprise	Local unit	Person	Ext_J	Industry	w_{Ind}	$Ext_J · w_{Ind}$
J1	EU1	LU11	**PIN1**	0.30	A	0.6	0.18
J1	EU1	LU11	**PIN1**	0.30	B	0.4	0.12
J2	EU1	LU11	PIN2	1.00	A	0.6	0.60
J2	EU1	LU11	PIN2	1.00	B	0.4	0.40
J3	EU2	LU21	PIN3	1.00	C	1	1.00
J4	EU2	LU21	PIN4	1.00	C	1	1.00
J5	EU2	LU22	PIN5	1.00	D	1	1.00
J6	EU2	LU22	**PIN1**	0.20	D	1	0.20
Total						6	4.5

Comments on the Job Register:

The register contains data on six jobs corresponding to 4.5 full-time jobs.

Jobs J1 and J2 are divided into two combination objects each, as LU11 is active in both Industry A and B. The weights 0.6 and 0.4 for these two Industries are taken from the *Local unit Register*. Ext_J refers to the extent of the work for each job.

Comments on the Employment Register:

If $Ext_J · w_{IND}$ is summed up for person PIN1 in the *Job Register*, the result obtained is 0.18 + 0.12 + 0.20 = 0.50. This value becomes Ext_P for PIN1 in the *Employment Register*.

Three combination objects for three Industries are formed for PIN1. Both Industry and job/local unit are multi-valued variables for persons.

The weights for combination objects are formed by multiplying w_{Ind} with w_{Job}, where w_{Ind} is taken from the *Job Register* and w_{Job} is calculated as every job's share of all jobs that the person has. For PIN1, job J1 has the weight (.18+.12)/(.18+.12+.20) = 0.6

3. Local unit Register				
Local unit	Industry	Ext_{LU}	w_{Ind}	$Ext_{LU} · w_{Ind}$
LU11	A	1.3	0.6	0.78
LU11	B	1.3	0.4	0.52
LU21	C	2.0	1	2.00
LU22	D	1.2	1	1.20
Total		3		4.5

4. Enterprise Register					
Enterprise	Local unit	Industry	Ext_{EU}	w_{Ind}	$Ext_{EU} · w_{Ind}$
EU1	LU11	A	1.3	0.6	0.78
EU1	LU11	B	1.3	0.4	0.52
EU2	LU21	C	3.2	0.625	2.00
EU2	LU22	D	3.2	0.375	1.20
Total			2		4.5

4. Consistent estimation of full-time employees by Industry

All four registers in Chart 9.23 give exactly the same estimates of how the volume of work, measured as number of full-time employees, is distributed between the different Industries. The reason for this is that all information of Industry is utilised when theoretically consistent weights are used in the four registers.

Table 1 in Chart 9.24 below compares different estimates of full-time employees by Industry. The columns 'Employment Register', …, 'Enterprise Register' are estimates that have been calculated the traditional way with the registers in Chart 9.22. The differences between these four columns are solely due to the estimation method being unsuitable. The last column in table 1 contains the estimates based on combination objects that have been carried out using the registers in Chart 9.23. Table 2 in Chart 9.24 below shows the traditional estimate formed using the register on persons in Chart 9.22 above and, in the right column, estimates are given where the weights w_P in the register on persons in Chart 9.23 above is used. For example, for Industry A, sum of w_P in the register on persons = 0.36 + 0.6 = 0.96 in table 2.

Chart 9.24 Comparison of different types of estimation

Table 1
Number of full-time employees by Industry

Industry	Traditional estimation, Chart 9.22				Estimation with combination objects Register 1–4 in Chart 9.23
	Employment Register	Job Register	Local unit Register	Enterprise Register	
A	1.5	1.3	1.3	1.3	0.78
B	0.0	0.0	0.0	0.0	0.52
C	2.0	2.0	2.0	3.2	2.00
D	1.0	1.2	1.2	0.0	1.20
Tot	4.5	4.5	4.5	4.5	4.5

Table 2
Number of persons by Industry

Industry	Traditional, Chart 9.22 Employment Register	With combination objects Employment Register in Chart 9.23
A	2	0.96
B	0	0.64
C	2	2.00
D	1	1.40
Tot	5	5.00

The traditional estimation method thus results in inconsistencies in the register-based statistics produced. Furthermore, differences in population and variable definitions should lead to further inconsistencies in real registers.

There is a fourth reason for why register-based statistics from different products can be inconsistent. The table above shows the effects of content-related differences. Table 1 describes *full-time employees*, while Table 2 describes *persons*. With statistics on persons and labour market statistics, it is common to describe persons but, with economic statistics, it is more common to measure volumes and full-time employees.

9.2.7 Multi-valued variables – what is done in practice?

Example: Change of Industry

Section 9.2.2 discusses the multi-valued variable 'industrial classification'. The traditional methodology means that all the activities in an enterprise with several industrial classifications are assigned to the largest industrial classification. This leads to level shifts in time series when the largest industrial classification for an enterprise changes. The size of the time series disturbance depends on how significant the enterprise is within the relevant industry or region.

The example below is based on reality but we have adapted the data slightly. During years 1–3, enterprise X Ltd has carried out activities within several industries, but around 60% of turnover relates to Industry R. During year 4, X Ltd bought another enterprise with activities in another industry. The change of ownership took place from quarter 4 of year 4 onwards.

Column (2) in Chart 9.25 below contains turnover in SEK millions for all enterprises in Industry R, excluding X Ltd. Columns (3) and (4) contain the total turnover for X Ltd. Column (5) has been taken from the Business Register and shows the share of the activities that is carried out in Industry R.

Using traditional estimation, the whole enterprise's turnover is allocated to Industry R during years 1–4. Note that Industry R remains as the principal industrial classification during the whole of year 4 as changes are only made at the turn of the year with the traditional methodology. From year 5, none of the enterprise's turnover is allocated to Industry R. By adding columns (2) and (7), we obtain the time series in column (8) that contains a time series level shift.

Chart 9.25 Estimation of turnover within one industry using two methods

Yr Q	Industry R excl. X Ltd	X Ltd before purchase	X Ltd after purchase	X Ltd's share within Industry R	Weight with traditional estimation	X Ltd's contribution to Industry R, traditional estimation	Industry R traditional estimation	X Ltd's contribution to R with combination objects	Industry R with combination objects
(1)	(2)	(3)	(4)	(5)	(6)	(7)	(8)	(9)	(10)
1 1	7 684	7 354		0.60	1	7 354	15 038	4 412	12 096
1 2	7 086	7 086		0.60	1	7 086	14 172	4 252	11 338
1 3	8 142	6 788		0.60	1	6 788	14 930	4 073	12 215
1 4	9 853	8 387		0.60	1	8 387	18 240	5 032	14 885
...	...	...		...	...	...	...	...	...
4 1	13 071	9 259		0.57	1	9 259	22 330	5 278	18 349
4 2	13 127	9 509		0.57	1	9 509	22 636	5 420	18 547
4 3	11 253	9 499		0.57	1	9 499	20 752	5 414	16 668
4 4	12 921		15 881	0.21	1	15 881	28 802	3 335	16 256
5 1	12 782		12 397	0.21	0	0 000	12 782	2 603	15 385
5 2	13 360		12 634	0.21	0	0 000	13 360	2 653	16 013
5 3	11 098		11 621	0.21	0	0 000	11 098	2 440	13 538
5 4	12 888		13 209	0.21	0	0 000	12 888	2 774	15 662

In accordance with the estimation methodology based on combination objects, the shares in column (5) should be used as weights when turnover for enterprise X is estimated. By multiplying the enterprise's turnover in columns (3) and (4) with the weights from column (5), we obtain the part of the turnover that relates to Industry R. By adding this part of the turnover to column (9), which shows the turnover for the other enterprises within Industry R, we get the time series in column (10), which describes the industry's turnover without any time series level shift.

The traditionally estimated turnover series is affected by a time series level shift caused by abruptly changed aggregation errors. For the period year 1 to year 4, the traditional method gives an overestimation of the industry's turnover. This error becomes even more significant during the fourth quarter of year 4. From year 5 onwards the turnover within Industry R is underestimated.

The series in column (10) has been estimated using the weights in column (5). Even if these weights are not perfect, they are considerably better than the traditional weights in column (6).

By using weights, errors can be considerably reduced and the quality of the time series increases substantially.

In Chart 9.26 the estimation methods are compared.

Chart 9.26 Turnover in an industry, two estimates

SEK billions per quarter

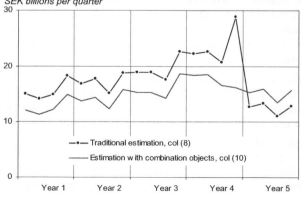

—•— Traditional estimation, col (8)

——— Estimation with combination objects, col (10)

Year 1   Year 2   Year 3   Year 4   Year 5

Multi-valued variables – summary of recommendations

A variety of important variables in the register system are multi-valued. The current way of handling these variables can, in some cases, produce estimates with aggregation errors. By using combination objects and weights when estimating, these errors can be reduced. In this section, a series of different estimation problems with multi-valued variables is described, and suggestions are made for solutions to these problems. The above example on the change in Industry shows how relatively simple methods, such as using weights, can bring about quality improvements, even though the weights being used are not completely perfect.

Another important advantage with the estimation method presented in this section (Section 9.2) is that economic statistics for different kinds of enterprise units can be made consistent with each other – these inconsistencies are today a serious problem.

9.2.8 Sample surveys with multi-valued register variables

When multi-valued variables from registers are used in sample surveys, the weights for the combination object in the register should be combined with the sampling weights. Totals should be calculated with all combination objects that belong to the corresponding sampled unit.

Example: In the Labour Force Survey, the interviewer asks about the local unit. The industrial classification of the local unit is then coded using information from the Business Register. Traditionally, only the principal industry is used which leads to aggregation errors in the Labour Force Survey. To avoid these aggregation errors the combination object *Local unit • Industry* in the register should be used in the data matrix of the sample survey.

The weights d_i and g_i are the common sampling weights and w_i is the weight for the multi-valued variable that describes the industry section in the Business Register.

$$\hat{Y} = \sum_{i=1}^{r_r} d_i g_i w_i y_i \qquad r_r \text{ is the number of responding combination objects} \qquad (5)$$

Suppose that a person *PIN3* has been selected and that this person works at local unit *LU4*, which is active within two industries, DJ and DH, with 60% in DJ and 40% in DH. The common sampling weight $d_i g_i$ is equal to 353.8 for this person.

Chart 9.27 Data matrix for the Labour Force Survey

Person	Gainfully employed	Hours worked	Local unit	Industry	$d_i \cdot g_i$	w_{IND}
Traditional estimation						
PIN3	Yes	36	LU4	DJ	353.8	-
Estimation with combination objects						
PIN3	Yes	36	LU4	DJ	353.8	0.6
PIN3	Yes	36	LU4	DH	353.8	0.4

How do the values of person *PIN3* contribute to these estimates?

With the *traditional estimation*:

– the number of employed persons within industry DJ increases by 353.8,

– the number of hours worked within DJ increases by 353.8 • 36 = 12 735.8 hours, and

– the industry DH does not increase, neither in number of employed persons nor number of hours worked.

With the *estimation method given in formula (5)* above:

– the number of employed persons within industrial classification DJ increases by
  $353.8 \cdot 0.6 = 212.3$,

– the number of employed persons within industrial classification DH increases by
  $353.8 \cdot 0.4 = 141.5$,

– the number of hours worked within DJ increases by $353.8 \cdot 0.6 \cdot 36 = 7\,642.1$ hours, and

– the number of hours worked within DH increases by $353.8 \cdot 0.4 \cdot 36 = 5\,094.7$ hours.

The sample surveys that use industry-specific estimates according to formula (5) have
fewer errors and sampling errors than the traditional estimates.

9.2.9   Combining collected and administrative data in a Business Register

Section 2.2.3 describes the different object types that are included in the Business Register.
Administrative data relating to legal units, LeU, and local units, LU, are delivered to Statistics Sweden.

To create statistically meaningful enterprise units, some legal units, LeU, must be combined
into enterprise units, EU and each EU consists of one or more local units LU.

Some EUs are divided into kind of activity units, KAUs, which are as *industry-specific* as
possible. In addition, some local units are divided into industry-specific local kind of activity units, LKAUs. Such divisions into industry-specific units are currently carried out only
when it is *possible to collect data* referring to these KAUs and LKAUs. During the work
with all these units, the statistical office as a rule contacts the enterprises concerned. Only a
limited number of large enterprises are contacted, and the number of EUs, KAUs and
LKAUs that are created is thus very limited.

Chart 9.28 Object types in the Business Register

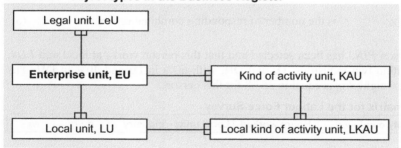

In Section 9.2, the concept *combination object* has been introduced for estimation purposes.
Estimation methods for the multi-valued variable *Industry* build on the fact that for every
combination of industry and enterprise unit or local unit, combination objects are created.
These combination objects are derived objects that have been created without mail questionnaires or telephone contacts with the enterprises concerned. These combination objects
are entirely industry-specific, but are not intended for data collection.

Some business surveys use a combination of collected questionnaire data from large enterprises and administrative data for the rest of the population. From the large enterprises,
industry-specific data are collected from kind of activity units (KAU) or local kind of
activity units (LKAU).

These collected data can be combined with administrative data for combination objects *Enterprise unit • Industry* or *Local unit • Industry* to generate industry-specific estimates for each industry. This is illustrated by the example below.

Example: Turnover survey combining collected and administrative data

The Business Register for a small region consists of five legal units LeU1–5. LeU1–3 belong to the same consolidated group, and are combined into one enterprise unit EU1. After discussions between representatives for the Business Register and this group, it is decided to divide the enterprise unit into two kind of activity units, KAU1 and KAU2. Questionnaires are sent to these two kind of activity units. For the rest of the population, the administrative units and administrative data are used.

Chart 9.29 A business survey with collected and administrative data

LeU-id	EU-id	KAU-id	Industry 1	%	Industry 2	%	Turnover	Source
LeU1–3	EU1	KAU1	DH	100			450	Collected data
LeU1–3	EU1	KAU2	DJ	100			300	Collected data
LeU4	EU2		DH	60	DJ	40	250	Administrative data
LeU5	EU3		DJ	70	DH	30	150	Administrative data

Traditional estimates of turnover by industry are:

– turnover DH = 450 + 250 = 700

– turnover DJ = 300 + 150 = 450

To reduce aggregation errors it is better to create combination units for each combination *Enterprise unit • Industry* for LeU4 (EU2) and LeU5 (EU3).

Chart 9.30 A business survey with collected and administrative data

LeU-id	EU-id	KAU-id	Industry	weight	Turnover	Source
LeU1–3	EU1	KAU1	DH	1	450	Collected data
LeU1–3	EU1	KAU2	DJ	1	300	Collected data
LeU4	EU2	KAU3	DH	0.6	250	Administrative data
LeU4	EU2	KAU4	DJ	0.4	250	Administrative data
LeU5	EU3	KAU5	DJ	0.7	150	Administrative data
LeU5	EU3	KUA6	DH	0.3	150	Administrative data
Total				4		

Estimates of turnover by industry, using combination objects are:

– turnover DH = 450 + 0.6 • 250 + 0.3 • 150 = 645

– turnover DJ = 300 + 0.4 • 250 + 0.7 • 150 = 505

Two combination objects are created respectively for LeU4 and LeU5. These four combination objects are four derived kind of activity units, KAU. Industry-specific estimates with smaller aggregation errors are calculated with the weights in Chart 9.30, where collected data always will have the weight 1.

9.3   LINKING OF TIME SERIES USING COMBINATION OBJECTS

Estimation methods for multi-valued variables are discussed in this chapter. When an object has several variable values for a multi-valued variable at the same time, a combination object is created for every combination of an object and a possible value. Every such combination object has a weight that is used when estimating values in the table cells. The same method can also be used for time series level shifts that are due to *changed classifications*, where the new and old categories are not entirely comparable. One advantage with the method presented here, which is based on the creation of combination objects, is that it is not necessary to first link the series at the macro level.

When for instance, the industrial classification system is changed, an old code can be replaced with one or many new codes. The old codes in earlier versions of the Business Register can as a rule be replaced with new codes if there is information about the product mix of the enterprises in the old version's register. Then the product mix can be translated into the new industrial codes. Bayard and Klimek (2003) have used this method, in combination with other methods, to link time series at the micro level. However, the method presented in this section needs less data and can be used when there is lack of information about the product mix.

When a classification system has been changed, the new and old codes can have the following relationships as shown in Chart 9.31 below.

Chart 9.31 Translation of old codes to new ones

Relationship between old and new codes	Old code	Code key	New code	Comments
One-to-one	1		A	No problems, the old code 1 is recoded to the new code A
Many-to-one	2 3		B	No problems, the old codes 2 and 3 are combined to the new code B
One-to-many	4		C D	Causes problems, how should the old code 4 be divided into the new codes C and D?
Many-to-many	5 6		E F	Same problem as mentioned above, how should: – old code 5 be divided up into E and F? – old code 6 be divided up into E and F?

One condition for the method we describe in this section is that there exists a register for a particular point in time or period in which every object is classified according to both the new and the old classifications.

If the classification is a stable characteristic, it is possible to recode to a new classification for identical objects in older register versions. But this type of reclassification can only be done for the part of the register population that can be traced back in time. When it is *not* possible to reclassify object-by-object in this way, the method using combination objects described below can be used.

Problems arise when objects with old codes are to be reclassified into one of several new codes. This is solved by combining every such object with every possible new code, and then these combination objects are given weights that are used when estimating the values in the table cells.

Example: Changed industrial classification code in the Business Register

In this example, we illustrate the method with fictitious data. The *industrial classification* is changed in year 3; the old codes 1 and 2 are replaced by the new codes A and B. For year 3, all enterprise units (EU) are coded with both the old and the new codes. For years 1 and 2, only the old Industry codes exist. Chart 9.32 describes how the old and the new codes correspond to each other.

Chart 9.32 Key between old and new industrial classification codes

Relationship between codes	Old code	Code key	New code	Comments
Many-to-many	1		A	The old 1s become the new A but the old 2s should be divided up into the new A and B
	2		B	

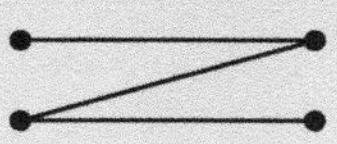

The old code 1 does not present any problems, as these should always be coded with the new code A. But the old code 2 causes problems: should the enterprise with the old code 2 be given the new code A or B? With the method using combination objects, two combination objects with codes A and B replace enterprises with code 2, respectively.

The chart below shows the Business Register for three years with data on the number of employees (*Nr of empl*) and *value added* for every enterprise unit. EU8 and EU13 have two industrial classifications, and have therefore two combination objects for which the weights are known.

Chart 9.33

Business Register year 1

Enterprise	Industry	Nr of empl	Value added	w_i
EU1	2	10	15	1
EU2	1	20	50	1
EU3	2	50	100	1
EU4	1	30	75	1
EU5	2	10	15	1
EU6	1	20	40	1
EU7	2	20	40	1
EU8	1	60	105	0.6
EU8	2	60	105	0.4
EU9	1	40	80	1
EU10	2	20	40	1
EU11	1	10	25	1
EU12	2	10	15	1

Business Register year 2

Enterprise	Industry	Nr of empl	Value added	w_i
EU1	2	10	15	1
EU2	1	20	50	1
EU3	2	50	100	1
EU13	1	40	80	0.5
EU13	2	40	80	0.5
EU14	1	10	25	1
EU15	2	20	40	1
EU16	1	40	80	1
EU17	2	30	60	1
EU18	1	40	80	1
EU19	2	20	40	1
EU20	1	10	25	1
EU21	2	10	20	1

Business Register year 3

Enterprise	Industry old	Industry new	Nr of empl	Value added	w_i
EU1	2	B	10	15	1
EU2	1	A	20	50	1
EU3	2	A	50	100	1
EU13	1	A	40	80	0.5
EU13	2	B	40	80	0.5
EU14	1	A	10	25	1
EU15	2	A	20	40	1
EU22	1	A	10	25	1
EU23	2	B	40	60	1
EU24	1	A	30	75	1
EU25	2	A	50	100	1
EU26	1	A	20	50	1
EU27	2	B	10	15	1

With the information in these three registers, it is possible to form time series according to the old industrial classification.

Values for year 3 can also be calculated according to the new industrial classification. As EU8 and EU13 conduct activities within two industrial classifications, the combination object *EU · Industry* is formed, giving the weights w_i that is used when the estimates for these time series are formed as described earlier in Section 9.2.

In the chart below, we can see the time series level shift for year 3, but how can the revised values for previous years be calculated?

Chart 9.34 Number of employees by Industrial classification years 1–3

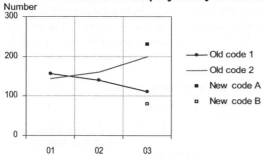

The chart below shows how the linking is carried out in this case. For year 3, we know both the old and the new Industry code for all enterprise units in the register. As industrial classification is a relatively stable variable, we use the new Industry code for year 3 also for years 1 and 2. In this way, EU1–3 receive the new Industry code for years 1 and 2, and EU13–15 receive the new codes for year 2. What should be done with the enterprises that are not in the register for year 3? The enterprises with the old Industry code 1 cause no problems and are given the new code A. These first easy steps in the linking process are illustrated below.

Chart 9.35 Linking at micro level – the easy steps

Business Register year 1

Enter-prise	Industry Old	New	Nr of empl	Value added	w_i old	w_i empl	w_i vadd
EU1	2	B	10	15	1	1	1
EU2	1	A	20	50	1	1	1
EU3	2	A	50	100	1	1	1
EU4	1	A	30	75	1	1	1
EU6	1	A	20	40	1	1	1
EU9	1	A	40	80	1	1	1
EU11	1	A	10	25	1	1	1

Business Register year 2

Enter-prise	Industry Old	New	Nr of empl	Value added	w_i old	w_i empl	w_i vadd
EU1	2	B	10	15	1	1	1
EU2	1	A	20	50	1	1	1
EU3	2	A	50	100	1	1	1
EU13	1	A	40	80	0.5	0.5	0.5
EU13	2	B	40	80	0.5	0.5	0.5
EU14	1	A	10	25	1	1	1
EU15	2	A	20	40	1	1	1
EU16	1	A	40	80	1	1	1
EU18	1	A	40	80	1	1	1
EU20	1	A	10	25	1	1	1

Business Register yr 3

Enter-prise	Industry Old	New	Nr of empl	Value added	w_i
EU1	2	B	10	15	1
EU2	1	A	20	50	1
EU3	2	A	50	100	1
EU13	1	A	40	80	0.5
EU13	2	B	40	80	0.5
EU14	1	A	10	25	1
EU15	2	A	20	40	1
EU22	1	A	10	25	1
EU23	2	B	40	60	1
EU24	1	A	30	75	1
EU25	2	A	50	100	1
EU26	1	A	20	50	1
EU27	2	B	10	15	1

What should be done with the enterprises that are not in the register for year 3 and have the old code 2? Because the old code 2 can become both the new codes A and B, combination objects are created for EU5, 7, 8, 10, 12, 17, 19 and 21.

How should we generate weights for these combination objects? These weights must show how the old code 2 is divided into the new codes A and B. For year 3 we can use available data in chart above to calculate such weights. We calculate different weights for the variables *number of employees* and *value added* as the new industries can have different structure. The calculations are shown in Chart 9.36 below. The values in extra bold type in the chart above are used.

Chart 9.36 Creating weights for the combination objects

Industry, old code	Industry, new code	Number of employees	Value added
2	A	50 + 20 + 50 = 120	100 + 40 + 100 = 240
2	B	10 + 40 • 0.5 + 40 +10 = 80	15 + 80 • 0.5 + 60 + 15 = 130
Total		200	370

- The weight w_i *old* produces estimates for the old Industry, while w_i *empl* and w_i *vadd* below produce estimates for the number of employees and value added according to the new industrial classification.

- Number of employees is 200, of which enterprises within Industry A have 120, so $120/200 = 0.6$ Then w_i *empl* is 0.6 for Industry A and 0.4 for Industry B

- Value added is 370, of which enterprises within Industry A have 240, so $240/370 = 0.65$ Then w_i *vadd* is 0.65 for Industry A and 0.35 for Industry B

We assume that these weights can be used for year 1 and 2, and in the chart below the combination objects with weights are shown.

Chart 9.37 Linking at micro level – the step using combination objects

Business Register year 1

Enter-prise	Industry Old	Industry New	Nr of empl	Value added	w_i old	w_i empl	w_i vadd
EU5	2	A	10	15	1	0.6	0.65
EU5	2	B	10	15	0	0.4	0.35
EU7	2	A	20	40	1	0.6	0.65
EU7	2	B	20	40	0	0.4	0.35
EU8	1	A	60	105	0.6	0.6	0.6
EU8	2	A	60	105	0.4	0.24	0.26
EU8	2	B	60	105	0	0.16	0.14
EU10	2	A	20	40	1	0.6	0.65
EU10	2	B	20	40	0	0.4	0.35
EU12	2	A	10	15	1	0.6	0.65
EU12	2	B	10	15	0	0.4	0.35

Business Register year 2

Enter-prise	Industry Old	Industry New	Nr of empl	Value added	w_i old	w_i empl	w_i vadd
EU17	2	A	30	60	1	0.6	0.65
EU17	2	B	30	60	0	0.4	0.35
EU19	2	A	20	40	1	0.6	0.65
EU19	2	B	20	40	0	0.4	0.35
EU21	2	A	10	20	1	0.6	0.65
EU21	2	B	10	20	0	0.4	0.35

Note: The weights for EU8 with industry 2 are calculated as follows (the white cells):

New A: 0.4 • 0.6 = 0.24 and 0.4 • 0.65 = 0.26
New B: 0.4 • 0.4 = 0.16 and 0.4 • 0.35 = 0.14

Combining data in Chart 9.35 and 9.37 into a complete registers, the linked time series values for *number of employees* and *value added* can be calculated for years 1 and 2. In the chart below, the time series relating to the old and the new classifications are compared.

Chart 9.38 Employees by industrial classification years 1–3

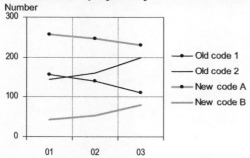

Number

- Old code 1
- Old code 2
- New code A
- New code B

In this example, the weights used for the calculations remain constant for all the years.

If the link concerns a new industry that has appeared during the last few years, the weights can be decrease back in time.

CHAPTER 10

Quality of Register-based Statistics

In *Statistics Denmark* (1995, Chapter 4) there is a summary of what is required of a register system so that it is possible to successfully use administrative data for statistical purposes. It states that register-based statistics will be of high quality if: *The register system has good coverage and it contains a wide variety of important object types, relations between objects and many variables.*

We mention in Section 1.6 that there are both prejudices and legitimate criticism against statistics based on administrative data. The prejudices consider register-based statistics to be cheap but of bad quality, compared to 'true' survey statistics. Where do these attitudes come from? Media articles such as the one in the example below, from the leading daily newspaper in Sweden, can often give rise to such attitudes.

Dagens Nyheter, 2 March 2003

Statistics lie about murder

A new computer system can lie behind figures that show three times the number of cases than there have actually been.

Mr. Mikael Rying has written a paper on the difference between the numbers of reported and statistically registered murders in Sweden and the number of actual murders.
– During the period 1990–1998, on average only 57 per cent of the reported cases were actual cases of murder and manslaughter, says Mikael Rying. The other cases related to everything from suicide and alcohol poisoning to simple coding errors in the police reports.

Year	Deaths by cause of death: Murder and manslaughter	Police reported crime: Murder and manslaughter
96	110	199
97	94	157
98	98	185
99	108	188
00	90	175

Average for 1996–2000:

	100	181

100/181 = 55%

Source: Statistical Yearbook of Sweden 2002–2003

This example shows that it is not a good idea only to use *one* administrative source. The fundamental idea behind the register system is that many sources should be used so that a high standard of quality and consistency can be achieved. Furthermore, administrative data should not be used directly, but should be edited and processed to suit the needs of the statistics. In this case, the problem with police reported crimes is that the reporting does not clearly differentiate between different object types: murder victims, suspected criminals and the murder itself. This can lead to coding errors.

Another administrative source that could be used is the Cause of Death Register, which is part of Statistics Sweden's register system and is checked against the Population Register. This source is not affected by the quality flaws that exist in the data regarding reported

---

crimes. In comparison, it is not likely that a sample survey would produce murder statistics of any better quality – a questionnaire to all or to a sample of the country's police stations would only lead to the double provision of largely incorrect data. Both the register-based surveys in this example should have easily accessible quality declarations so that users do not incorrectly interpret the data.

Another common opinion of administrative data is that the respondents only give the data that helps their purposes. In the example below, it is a fact that most people would like to pay as little tax as possible, so there is a possibility that the deductions may be higher than is justifiable. (From the Swedish newspaper Dagens Nyheter, 16 July 2003)

---

80 per cent of Swedish people's tax deductions are pure tax evasion

Taxpayers submit errors worth billions in their tax declarations. Complicated rules and unclear legislation have made it hard for the country's tax authorities to check all the deductions. Errors can be found primarily in the deductions for share transactions, management fees and other share-related charges.

Deductions for the sales of shares
– 1/3 of all share sales contain errors
– 700 000 taxpayers report profit of around SEK 50 billion and losses of around SEK 10 billion
– Tax errors are difficult to judge and amount to billions of Swedish kronor
– Many inadvertent errors occur because of the complicated rules

Deductions for management fees
– 125 000 taxpayers claimed deductions of a total of SEK 515 million
– 66% of these deductions contain incorrect information
– Tax errors can in total be calculated at SEK 90 million
– A deduction for fees for fund managers is the most common error, the fee is deducted automatically

Deductions for other expenditure
– 700 000 taxpayers claimed deductions of a total of SEK 2.8 billion
– 82% of these deductions contain incorrect information
– Tax errors can in total be calculated to amount to around SEK 700 million

---

The headline exaggerates in several ways, '80 per cent' is an exaggeration and 'tax evasion' is often based on misunderstanding due to complicated rules:

– Deductions for share sales: the errors are largely unintentional due to complicated rules.

– Deductions for management fees: the errors are on average 17% (90/515) and the most common error can be unintentional due to misunderstanding.

– Deductions for other expenditure: 82% of these deductions contain errors but the deductions are on average 25% (700/2800) incorrect.

Another perspective on these errors is when they are compared to the total income for all those filing tax declarations, and then the error is of the size 0.3%.

The fact that deductions in the declarations are too high, and that consequently the tax is too low, does not mean that the statistics in the Income Register are of low quality, despite the fact that they are based on these declarations. Suppose that we have data for a person who shows too high deductions on her/his tax declaration, but otherwise declares correctly:

Income from employment	257 600	The income is correct
Deductions for other expenditure	25 500	The deduction is too high but is accepted
Taxable income	232 100	Taxable income is incorrect according to the tax rules but is not used for the statistics
Tax	100 000	The tax is incorrect and too low according to the tax rules, but statistically it is correct, as this is the tax that the person actually paid
Disposable income	157 600	Statistically correct

Statistics Sweden's statistics regarding earned income are not incorrect because of this person's data; neither are the statistics regarding disposable income incorrect, as this is formed by calculating the difference between income and tax actually paid.

10.1   SPECIFIC QUALITY ISSUES FOR REGISTER-BASED STATISTICS?

What is statistical quality when it comes to register-based statistics? Platek and Särndal (2001) state that the statisticians lag behind in building a theory of accuracy assessment for statistics based on registers, and that a theory is needed.

Holt (2001) points out that there are important differences between statistics from surveys with their own data collection on the one hand, and statistics from administrative sources on the other. Holt maintains that the most important aspect of quality when it comes to register-based statistics is not accuracy, but relevance.

Nanopoulos (2001) maintains that countries like Denmark that have well integrated register systems need a conceptual apparatus regarding errors in statistics, which will be different from that required by countries that mainly carry out sample surveys and censuses.

Our conclusion is that it is important to consider the following when discussing the quality of register statistics:

– It is necessary to distinguish between surveys with their own data collection and ones that are register-based. Otherwise, there is a risk of uncritically using the traditional error models developed for sample surveys and censuses.

– There should also be a distinction between the quality of a register-based survey and that of a statistical register, as a register has many possible uses.

Sample surveys and censuses are carried out with *one* particular use in mind and quality issues generally focus on the estimates carried out. In the case of a statistical register, *many* different uses are possible – such a register may serve not only current surveys but also future ones.

The chart below shows the difference between the register and the register-based survey. Further, it shows the two aspects that are crucial with regard to the quality of the register: the administrative systems and the system of statistical registers. These are the two sources on which the statistical register is based.

Chart 10.1   Quality in a register and a register-based survey

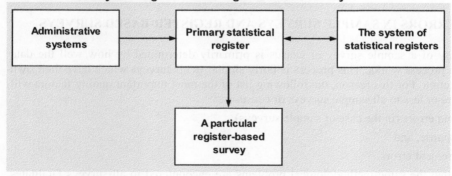

Similar to other surveys, the quality of a register-based survey also relates to *one* specific use of the register and also focuses on the quality of the estimates, particularly their relevance and accuracy in relation to the purpose of the survey. Describing quality is here a question of indicating whether the quality of the survey is *good or bad*.

However, the quality of the statistical register itself is not related to one particular use and, when describing quality in this respect, it is important to indicate *what characteristics* the register has, thereby implying the uses to which it may be put. The quality of the register will affect the quality of the surveys based upon it, and is determined by three factors:

– the administrative systems on which the register is based,
– the possibilities offered by the system of statistical registers with regard to improving coverage, content of variables and consistency, and
– the processing done to produce the register.

The first factor that determines the quality of the register is the *administrative systems* upon which it is based (see Chart 10.1). Administrative systems are generally unique – the administrative data collection in the case of population registration, for instance, is altogether different from that for enterprises' statements of earnings. Data collection within the administrative systems is also usually different from the data collection of a statistical office. Even though the administrative systems involve respondents filling in forms, the reporting of information has its own conditions, and is governed by administrative rules and regulations.

The second factor that determines the quality of the register relates to the possibilities that the *register system* offers. The register should be co-ordinated with the rest of the register system, and the system as a whole should function efficiently.

The third factor that determines the quality of the register is the *processing* performed when the register was created. How was the register population defined, how was the content of variables determined and how were the data edited?

The administrative system and the internal processing of data have hitherto been perceived as specific to each register, which has hampered the exchange of experience and the development of methodology, but the quality of one particular register affects other register-based surveys that have a use for that register. Thus, for example, missing values of the variable *industrial classification* in the Business Register constitutes a problem for all the registers and surveys that include this variable. Furthermore, all others are dependent on this problem being solved by those responsible for the Business Register, because otherwise statistics from different registers will not be consistent.

10.2   ERRORS IN SAMPLE SURVEYS AND REGISTER-BASED SURVEYS

The quality of a sample survey or census is primarily determined by how well the data collection process works. This process is fairly similar in all surveys which have their own data collection. For this reason, the following list of the most important quality factors will apply more or less to all sample surveys or censuses:

– sampling errors (in the case of sample surveys),
– nonresponse, and
– measurement errors.

The fact that the same methodological problems are encountered in all surveys facilitates discussions with colleagues, methodology development and the establishment of guidelines. On the other hand, the different sample surveys do not affect one another – nonresponse in the Labour Force Survey, for instance, does not affect, say, the Living Conditions Survey or the survey on Deliveries and New Orders in Industry.

Although the most important quality factors will probably not be the same for all register-based surveys, the quality of one register will, in general, affect the quality of others. Surveys based on data collection and register-based surveys can thus be compared in the following way:

Surveys based on data collection:	Register-based surveys:
– Same quality issues in all surveys	– Different quality issues in different surveys
– Quality of one survey does not affect other surveys	– Quality of one survey affects many other surveys

Until now, there has not been much exchange of experience concerning register-statistical methodology and quality issues – it is our hope, however, that a common terminology and a common perspective will stimulate such exchange.

10.2.1 Sampling errors and integration errors

Sampling errors have for a long time been regarded as the most important error in sample surveys. Sampling designs and estimation methods have therefore been developed to reduce this kind or error. In the book by Cochran (1963) twelve out of thirteen chapters are devoted to these issues, in the last chapter, measurement errors and nonresponse are mentioned.

In register-based surveys there is no sampling phase; instead this kind of survey is dominated by the *integration phase*, where data from different sources are integrated into a new statistical register. During the integration phase the register population and derived objects are created, variables are imported from different sources and derived variables are created. The kinds of errors that have their origin from the integration phase should be called *integration errors*. In this category coverage errors, matching errors, missing values due to mismatch (discussed in Section 8.1) and aggregation errors (discussed in Sections 9.1 and 9.2) are included. In Section 10.4, we present 40 quality indicators for register-based surveys; 22 of these have their origin from the integration phase.

10.2.2 Errors in detailed tables

For sample surveys, the sampling error determines how detailed a published table can be. But for register-based surveys there is no sampling error that can tell if a table is too detailed to be published. In register-based surveys there are other kinds of errors that are small at a more aggregate level, but which become disturbing if too detailed tables are published.

Hierarchic classifications
Important statistical standards or classifications as Occupation, Education and Industry, contain hundreds or thousands of categories at the most detailed level. However, in many cases, the quality is not good at the most detailed level. With sample surveys, the sampling error only permits that estimates on more aggregate levels are published. With register-based surveys and censuses, other methods must be used to prevent that tables with low quality are used or published.

The solution of this problem has been mentioned in Section 2.5, where the concept *standardised variable* is introduced. The important classifications should be defined as standardised variables, and the registers, where these variables are created for the first time in the register system, are responsible of documentation and quality.

For instance, those responsible of the Business Register should examine the quality of the variable *Industry*. They should also group the detailed categories of this variable and create a derived variable, which is suitable for register-based tables. All other register units should then use this derived variable in their published tables.

Model errors

When the values of derived variables are estimated with a model, as in Section 6.2.3, model errors can be regarded as random. The model errors should be examined with special surveys and the results from these surveys can be used to estimate systematic and random model errors.

In Section 6.2.3 the table to the left in Chart 10.2 is given. In the Labour Force Survey (LFS) 31 458 persons have been interviewed and classified as employed or not employed. If we trust in the quality of the LFS, the probabilities of the two kinds of model or classification errors of the derived variable *Employed* in the Employment Register can be estimated as 5%, respectively 13%.

Chart 10.2 Classification errors in the Employment Register 1993

Number of persons in test data	Estimate in Employment Register			Estimated classification errors:		
	Employed	Not employed	Total	Employed	Not employed	Total
Employed LFS	22 360	1 158	23 518	95%	**5%**	100%
Not employed LFS	1 068	6 872	7 940	**13%**	87%	100%
Total	23 428	8 030	31 458			

If we assume that all employed and unemployed are classified with these risks of errors in the register, we can estimate quality components in the following way, where the number of persons classified as employed is the sum of two independent stochastic variables with binomial distributions.

In the chart below, the true numbers of employed and not employed persons in two domains are compared with the corresponding expected numbers and standard errors based on the estimated classification errors in Chart 10.2.

Chart 10.3 Quality of estimates for two domains in the Employment Register

	True number of:		Nr of persons	**Expected** number of	Standard error of
Domain	Employed	Not employed	in domain	employed in the register	the register estimate
1	100	100	200	95 + 13 = 108	4.0
2	50	10	60	47.5 + 1.3 = 48.8	1.9

An estimated employment rate of 54% (=108/200) in domain 1 with 200 persons, is expected to have a systematic error of approximately 8/200 = 4 percentage points, with a standard error of 4/200 = 2 percentage points.

An estimated employment rate of 81.3% (=48.8/60) in domain 2 with 60 persons, is expected to have a systematic error of approximately –1.2/60 = –2 percentage points, with a standard error of 1.9/60 = 3 percentage points.

The information about the systematic error could be used to produce adjusted estimates, and the information about the standard errors could be used to indicate that table cells are based on too few observations.

The general conclusion of the discussion in this section is that the quality of the variables in the register must be known. Those responsible for the register must understand how detailed tables can be. In Section 8.2 the overcoverage in the Swedish Population Register is mentioned. The overall undercoverage is very small, but for some categories born outside Sweden the undercoverage will disturb detailed tables. This quality information must be known by all those who use data from the Population Register. Some variables suffer from missing values. Imputation errors will cause random errors that will disturb the estimates in too detailed tables. Also, the extent of missing values must be known by all those that use these variables.

10.2.3 Random variation in register data

When interpreting estimates from sample surveys, you should always ask the following questions:

– Are differences or changes significant from a subject matter perspective? The user should be able to answer this question without support.

– Are differences or changes statistically significant as compared with standard errors? The users need help with this judgement.

The first question should also be asked when interpreting estimates from register-based surveys, but there is no similar tradition corresponding to question 2 above. There is a risk that users over-interpret estimates from censuses and register-based surveys – as there is no sampling error everything is statistically significant! To prevent such interpretation of register-based statistics, differences and changes should be judged against some other measure of randomness in data than the sampling error. As most register-based surveys are yearly or monthly surveys, we suggest that the time-series noise is used for this purpose.

With software for time-series analysis, a monthly series can be decomposed into trend, seasonality and noise. A yearly series can in the same way be decomposed into trend and noise. Even when there is no sampling error, the time-series noise can be substantial. In time-series analysis it is customary to interpret this as *natural random variation.*

The charts below present actual frequencies of accidents. Short-term variations in such frequencies must not be interpreted as indicating that the underlying accident risks have changed. In small regions with few children, the natural random variation is extremely marked, whilst in large regions with a lot of children it will be less.

Chart 10.4   Road accidents, boys 0–12 years, actual values and estimated trends
A. County with roughly 5 000 boys   B. County with roughly 125 000 boys

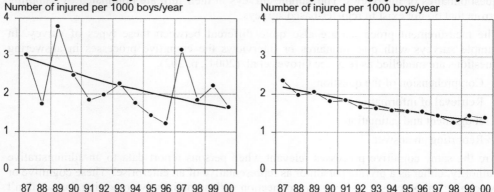

In Chart C and D below two other kinds of tend patterns are shown. The trends in these charts have been chosen so that the time-series residuals (the 'noise') are not autocorrelated and that the standard error of the residuals is close to the theoretical standard deviation according to the Poisson distribution.

Chart 10.4 continued

C. County with no trend
Number of injured per 1000 boys/year

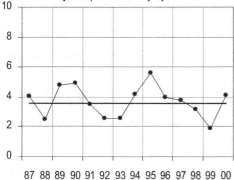

D. County with moving average trend
Number of injured per 1000 boys/year

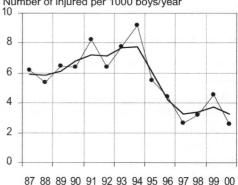

The series in Chart A–D have been created with register-based statistics from the Swedish Patient Register and the Population Register. The message to the users is clear:

In Chart C there are no statistically significant changes, but in Chart D the accident risk increased during 87–94 but after that the risk has been decreasing.

Our suggestion is the following: monthly, quarterly or yearly time-series from register-based surveys can be analysed, and the standard error of the time-series residuals can be used to judge if differences and changes are statistically significant.

10.2.4   Measurement errors

There are many important differences between error structures found in statistical registers based on administrative records compared to data collected in sample surveys. We have already mentioned that integration errors are important in register-based surveys and that these kinds of errors do not exist in sample surveys where all variables are collected via questionnaires or interviews. The sample surveys in their turn are dominated by sampling errors that do not exist in register-based surveys.

The measurement processes are also quite different between these types of survey. In sample surveys with questionnaires or interviews the cognitive processes in answering questions are modelled as (e.g. see Groves et al. (2004), p. 202):

– Comprehension of the question

– Retrieval of information

– Judgement and estimation

– Reporting an answer

Are the same cognitive processes relevant when persons report data to an administrative authority, either as a private person or as representative of an enterprise? These cognitive or psychological processes also exist in connection with administrative reporting, but we don't think they are so important. Instead, administrative rules and legislation are important factors, and when reporting data from enterprises, accounting principles and practise are more important than psychology.

In the chart below the two ways of collecting data are compared:

Chart 10.5 Measurement errors – comparison of data collection methods

Collecting data in sample surveys	Collecting data in administrative systems
Underlying structure of question: *Will you please try to understand our questions and try to remember?* *It is not necessary that you answer, and it does not matter what you answer, we will not do you any harm*	Underlying structure of question: *1. Report last month's turnover before the 12th this month!* *2. Pay 25% of reported turnover before the 12th this month!* *3. If you don't report and pay, you will have to pay extra!*
Questionnaire to persons: Does the right person in the household answer?	Reports to authorities from persons: In our yearly tax form, we only add our signature and perhaps try some deductions
Questionnaire to enterprises: Does the right person within the enterprise answer?	Reports to authorities from enterprises: Regular duty of professionals, the enterprise's accounting system as a rule generates the report Errors are errors in the accounting routines or typing or scanning errors
Interviewer effects can be disturbing	No interviewers, no interviewer effects
Leading questions in market research is often a problem	Legally complicated questions
Variables collected via questionnaire or interview in sample surveys (or censuses): Measurement errors are important	Statistical variables in register-based surveys are often derived variables based on administrative variables: Relevance errors and model errors are important

It should be noted that some questions or variables collected in an administrative system are legally important, when other questions are not important. The quality of these unimportant questions can be lower – you can answer what you want, it will not hurt you; the preconditions are the same as for a question in a sample survey.

10.3   THE USERS' AND THE PRODUCERS' VIEW OF QUALITY

How are register-based statistics used? What are the demands of the users regarding quality? Biemer and Lyberg (2003) discuss quality in sample surveys. The starting-point is *one* estimate, and the total error for this estimate is divided into 12 components. Errors that occur during the different stages of a sample survey can be either random or systematic. *Random errors* make the estimates uncertain but do not cause distortion. On the other hand *systematic errors* do cause distortion, which is to say that the value sought is overestimated or underestimated.

Chart 10.6 Risk of random and systematic errors by major error source

Error source	Risk of random errors	Risk of systematic errors
Specification error	Low	High
Frame error	Low	High
Nonresponse error	Low	High
Measurement error	High	High
Data processing error	High	High
Sampling error	High	Low

Source: Biemer and Lyberg (2003, p. 59)

This is the *producer's* view of quality. For the producer of statistics it is important to know which parts of the survey function well, and which parts function badly. On the basis of this knowledge the processes where the most serious errors occur can be improved. We shall discuss here corresponding error components that occur in register-based surveys.

Platek and Särndal (2001) discuss the quality of official statistics from the *user's* point of view. The user is interested in the answers to questions as, 'Can I trust these statistics?' and 'Are they suitable for my purposes?' Users want a guarantee of quality. Can the statisticians give it? What form should it assume?

The answers to these questions given by statistical offices are often insufficiently clear. Platek and Särndal claim that data quality means different things to different categories of staff in a statistical office:

– statistical methodologists regard it as a question of accuracy,

– subject matter specialists regard it a question of content and presentation,

– informatics specialists regard it as a question of the efficient functioning of data systems and processing, and

– managers regard it as a question of the functioning of budgets and time plans.

From detailed knowledge on quality to a comprehensive picture of quality
The discussion in this section focuses on the gap between in-depth knowledge and a comprehensive overview. The different categories of staff have thorough in-depth knowledge of a great many different factors that affect quality. The methodologist thinks in terms of the different phases of the survey and all the sources of error that may exist in these phases, while the IT specialist thinks of the production system and all the processing errors that can occur. Both have extensive knowledge, but may lack the comprehensive overview that the user needs.

The subject matter specialists generally have the closest contact with the users of the statistics, which is why they are the ones who should provide this comprehensive overview of quality. However, to do so they must be in close contact with the methodologists and the IT specialists as well as being able to understand the users' needs.

Detailed knowledge of the different quality components is acquired using the quality assurance guidelines in relation to the processes involved to produce the statistical registers. On the basis of this knowledge, an overall appraisal of the quality of the register and the register-based survey can be made. Based on our conclusions above, we will discuss quality of register-based statistics in two ways:

– First we describe quality indicators for the different steps in a register-based survey. These indicators should be used by the producer of register-based statistics for quality assurance of the different steps of the production process.

– Finally we describe overall quality appraisals of quality that can be understood by the users of the register-based statistics.

10.4   DETAILED KNOWLEDGE OF A REGISTER'S CHARACTERISTICS

The purpose of quality assurance is to investigate and remedy quality defects, which can occur in different parts of the work on a register. The traditional error model follows the different steps of a sample survey. In corresponding fashion, we describe the quality of register-based statistics with the aid of groups of indicators for the different parts of the work of creating a register:

Chart 10.7 The steps to create a register – corresponding quality characteristics

Steps to create a register	Quality characteristics of a register
1. Determining the research objectives 2. The inventory phase 3. The planning phase	Group 1: Subject matter competence and capacity for development
4. Supplier contacts, receipt of administrative data	Group 2: Supplier contacts and editing
5a. The integration phase – the object set	Group 3: Characteristics of the object set, both in a base register and in general
5b. The integration phase – the object type	Group 4: Characteristics of the definition of objects
5c. The integration phase – the variables	Group 5: Characteristics of the variables
6. Documentation	Group 6: Characteristics of the documentation

10.4.1 The phase of determining research objectives – its effect on the register

What statistical needs and requirements is the register supposed to fulfil? Have contacts been established with advanced users or researchers? Will the statistical office carry out its own advanced analyses and reports? If it is known what is required of a register, it is then possible to build up an understanding of the uses to which the register may be put. Advanced users of statistics are an important group when looking at the development and application of a register. They often have valuable experiences and ideas which ought to be documented.

10.4.2 The inventory phase – how has it affected the register?

The inventory phase looks at the different sources that have been used to produce the new register. This can consist of administrative sources and statistical registers that already exist within the system. Do these sources, on the whole, have a rich content? If several sources have been integrated into the register, this will be of advantage to new users. The inventory should also investigate whether there are other sources connected to the area of study that have not been used. This can be an opportunity to carry out an active search for new sources.

10.4.3 Are any changes planned?

Are there plans for any changes or improvements regarding the register population, the object definitions or the content of variables? If so, it is a sign both of defects in the present register and of the existence of development work designed to increase the usability of the register. Sections 10.4.1–10.4.3 can be summarised as in the following chart:

Group 1. Subject matter competence and capacity for development

Quality indicator	Implies	How is this measured/appraised?	See section
1a Good contacts with users	Subject matter competence		10.4.1
1b Perform own analyses			
1c Sources are integrated	Diversified use intended	Qualitative appraisal	10.4.2
1d Search for new sources	Capacity for development		10.4.3
1e Changes are planned			

10.4.4   Supplier contacts and editing

How does contact with the supplier occur? These contacts should be deep enough for the supplier to understand the needs of statistics and for the recipient to understand the conditions governing the administrative system. The contacts should be regularly renewed, so that the recipient gets information about coming changes.

How are data and metadata received from the supplier? The editing of these data should be examined: is there only a simple editing of each administrative source or is there also consistency editing where several sources are compared? The experiences that are gained through the data editing process can also be analysed. Is the frequency of the different types of error documented? How is this information fed back to the supplier? These questions are taken up in the following chart:

Group 2. Indicators for how supplier contacts and editing function

Quality indicator	Implies	How is this measured/appraised?	See section
2a Group from the statistical office regularly in touch with supplier or one person occasionally in contact	Great or restricted subject matter expertise respectively	Qualitative appraisal	6.3.1 6.3.5
2b Invalid records and duplicates discovered during editing	Quality flaws in the administrative source	Number/proportion of invalid items and duplicates	
2c Differences between register population and base register discovered	Undercoverage Overcoverage	Number/proportion of missing and alien objects	
2d False hits discovered during editing	Links indicate wrong relation or wrong identity	Number/proportion of false hits	6.3.4
2e Missing objects and/or variable values discovered during editing	Incomplete delivery	Number/proportion of missing objects/variable values	
2f Obviously wrong variable values discovered during editing	Quality defects in administrative source	Number/proportion and size of different types of errors	
2g Both usual editing of each source and consistency editing comparing many sources	Diversified use intended	Qualitative appraisal	6.3

10.4.5   The integration phase – how has the object set been created?

In this phase, it should be analysed how existing sources have been processed to ensure that the new register contains the desired object set. This involve the editing of administrative data, the matching of different sources and selection of objects or the processing of time references to produce the object set for the designated point in time or period.

The following two charts show indicators that refer to base registers and those that can be applied generally to statistical registers.

Group 3.1 Indicators for the characteristics of the object set in a base register

Quality indicator	Implies	How is this measured/appraised?	See
3.1a Delay in reporting	Overcoverage and undercoverage in current version of register give errors in cross-sectional data and errors changing over time	Compare present version of register with version based on all available sources when all information is available. Indicate the size of revisions and length of delay. Indicate frequency of changes.	See below
3.1b Defects in reporting relating to objects that have ceased to exist	Overcoverage, giving errors in cross-sectional data and errors changing over time	This can cause implausible values in tables. Use existing surveys with their own data collection: returned mail, no contact for telephone interviews. Compare with other register information. Estimate the extent of overcoverage.	8.2
3.1c Objects with wrong information	Overcoverage and undercoverage	Investigate number/proportion of wrong notification data corrected or annulled	See below

Delay in reporting
With regard to the Swedish Population Register the intention is to continuously follow different monthly quality indicators in respect of delays in the registration of births, deaths and changes of address. Different local tax authorities can have different patterns of delay.

With regard to the Business Register the intention is on the one hand to follow frequencies of change per month in respect of newly registered and deregistered businesses, together with such frequencies in respect of address and industry, on the other hand to follow how up-to-date this information is.

Wrong information about objects
When it comes to the Swedish Population Register there is a special quality database where all incorrect notification information is saved which has later been rectified or annulled by the National Tax Board.

Group 3.2 Indicators of the characteristics of the object set in general

Quality indicator	Implies	How is this measured/appraised?	See
3.2a Register population not in agreement with base register	Lack of consistency and coherence	Match register with base register, analyse mismatch	5.4.4, 5.4.5 Chapter 13 Below
3.2b Administrative source contains objects not in base register	Undercoverage in base register	Analyse extent and cause of mismatch, report to base register	5.4.7
3.2c Base register contains objects not in source	Undercoverage in source register or overcoverage in base register	Analyse extent and cause of mismatch	5.5.7
3.2d Matching errors – false hits	Wrong relation, objects do not belong together despite having same values on linkage variables	Compare variables from integrated register	5.4.7 5.5.6 6.3.2

The register population is not in agreement with the base register
One example of inconsistent populations, which we found when we studied the register system of Statistics Sweden, can be seen in agricultural statistics, where the agricultural enterprises were not identical to those found in the Business Register. The same was true of

the educational and energy enterprises. A further example is provided by accommodation statistics, where the enterprise population was put together independently of the Business Register.

10.4.6   The integration phase – how have the objects been defined?

What processing has been done in order to check and change the object definitions? For example, have the administrative data been checked and adjusted so that the definitions are those that are required? Have derived objects been formed in the new register? Is the quality of the object definitions checked? Are register maintenance questionnaires or evaluations conducted? How large are the errors?

Group 4. Indicators for the characteristics of object definitions

Quality indicator	Implies	How is this measured/appraised?	See
4a Object definition deviates from norm	Lack of consistency and coherence	Qualitative appraisal	Below
4b Object definition not comparable over time	Lack of consistency and coherence	Qualitative appraisal	6.4
4c Are register maintenance questionnaires conducted?	Quality awareness	How often are such questionnaires conducted, and regarding which categories of objects?	5.4.6–7 5.5.3 6.3.2

Object definition deviates from the norm
In the Swedish Real Estate Assessment Register, objects (and also variables) are defined in administrative terms rather than in coordination with other Statistics Sweden surveys; the term 'block of flats', for instance, is defined differently. Similarly, the definition of a family in taxation terms in the Population Register is not in agreement with the household definitions in other Statistics Sweden surveys.

10.4.7   The integration phase – how have the variables been created?

This phase should look at what processing has been done to produce the intended variables. This can include looking at whether the variables in the administrative sources have been edited and also at the scope and treatment of missing values. The various sources, from which the variables have been imported, should also be documented.

The scope of any possible errors should also be investigated, such as measurement errors or errors of classification in the spanning variables. The methods used to detect errors could include sample surveys or special evaluation surveys. Focus groups and cognitive interviews analysing the administrative forms could also be conducted to discover sources of measurement error.

Group 5. Indicators for the characteristics of the variables

Quality indicator	Implies	How is this measured/appraised?	See
5a Matching errors – missing hits	Error in linkage variable	Identical objects have different values for linkage variables	Below
5b Matching errors – false hits despite same values for linkage variables	Error in linkage variable	Wrong relation, objects do not belong together despite same values for linkage variables. Compare variables from integrated register – implausible relations indicate errors	5.4.7 5.5.5 5.5.6 6.3.2
5c Variable definition deviates from norm	Lack of consistency and coherence	Qualitative appraisal	Below
5d Missing values and imputation errors	Quality defects of administrative source	Extent and causes of missing values are given for every variable	6.1.2 6.3 8.1
5e Delay in reporting	Revision error	Compare preliminary estimates with definite ones	Below
5f Wrong preliminary values	Revision error	Investigate number/proportion of wrong records which have been rectified or annulled	Below
5g Variable not used by supplier	Measurement error	Investigate whether the variable is used and edited by supplier	Below
5h Suspected low/high estimates	Measurement error	Such errors can be appraised by means of sample surveys and cognitive interviews	Below
5i Coding errors	Classification error	Same data are coded independently again, permitting the estimation of coding errors	6.2.4
5j Aggregation errors in multi-valued variable	Error in cross-sectional data and over time	Register contains only the most important value for the variable	9.2
5k Classification error in spanning variables	Structural error, relationship between variables weakened	Quality control surveys. The errors can be appraised and estimated by means of special sample surveys	Below
5m Model error in derived variable	Measurement error or classification error	The errors can be appraised and estimated by means of ordinary or special sample surveys	6.2.3

Matching error – missing hits
The addresses of the local units form a link between the Swedish Business Register and Real Estate Register. An attempt to match local units using this link did not succeed in about 30% of cases of persons employed in the municipality where the matching was least successful. By editing all local units with at least 25 employees, this percentage could be reduced to about 10%.

The variable definition deviates from the norm
The definitions in the Swedish Foreign Trade Register are not the same as those in the National Accounts and in the Structural Business Survey.

Delay in reporting (i)
Delays in the monthly reporting of Swedish housing construction cause underestimation of the number of apartments under construction. The preliminary estimates will later be re-

vised. It is important to follow the length of the delay and to make up-to-date estimates of the revision error, including both its systematic and its random components.

Delay in reporting (ii)
Delays in the monthly reporting of wage sums, employment tax, etc. give rise to nonresponse. The preliminary estimates are based on imputed values and will be revised when the information is received.

Wrong preliminary values
The information from tax payers may later be changed either because the tax payers have themselves indicated such change or because the tax authorities have done so.

Not used by the data supplier
On the Swedish statement of earnings, the employer should indicate the first and last month of the employment. The tax authorities have no use of this information, for which reason it is of poor quality.

Suspected low/high estimates
The owners of single-dwelling houses in Sweden tend to underestimate the residential floor space when it comes to tax appraisal, because this can mean lower real-estate tax.

Classification errors in the spanning variable
The structure of the population, or the distribution according to the spanning variable (e.g. industry), is incorrect. Correlations are weakened and inter-category comparisons are disturbed. Suppose, for instance, that a comparison is to be done between manufacturing and service enterprises from the point-of-view of profitability. With no classification error in the case of industry, the comparison will be the true one shown in Chart 10.8 whilst, with a classification error of 10%, the comparison will be as shown in Chart 10.9.

Chart 10.8 Without classification error

Industry	Turnover	Profit	Profitability
Manufacturing	300	30	10%
Service	250	10	4%

Chart 10.9 With classification error

Turnover	Profit	Profitability
300–30+25=295	30–3+1=28	9,5%
250–25+30=255	10–1+3=12	4,7%

10.4.8   Documentation as a part of quality assurance

Documentation work is also an important part of quality assurance. Incorrect and uncritical use of administrative data can be prevented using metadata that provide information about problems of comparability. Changes in the administrative system cause these kinds of problems, and must therefore be documented. There is otherwise a risk of arriving at incorrect conclusions.

Since it is possible for a statistical register to be used by various users working with the register system, the documentation of registers must be done in such a way as to make it accessible to all.

Metadata play a major role in the work on register-based statistics. In the case of the integration of different registers, it is necessary to know the definitions, and what the comparability problems are. It is important that the processing methods should be documented as well, to facilitate the development of methodology and the exchange of experience.

Documentation of registers and register-based surveys are discussed in Chapter 11. A good documentation should give answers to the questions in the chart below.

Group 6. Indicators for the characteristics of the documentation

Quality indicator	Implies	What should be documented?	See
6a Definition of register population	Content of statistics	Is the definition complete? Type of object, which objects are included, time references, geographical definition, relation of objects to area. Which sources have been integrated? Which processing methods have been used to create the register population?	5.4.1
6b Definition of variables	Content of statistics	Is the definition complete? Type of object, time reference of variable, measurement method, scale. Which sources have been integrated? Which processing methods have been used to create the variables of the register?	6.1.1
6c Population comparable over time	Time series quality	Time series level shifts and their effects. What processing has been done to provide comparability over time?	8.3 9.3
6d Variables comparable over time	Time series quality	Time series level shifts and their effects. What processing has been done to provide comparability over time?	8.3 9.3
6e Population's objects comparable over time	Longitudinal quality	Do the objects have stable identities? Can they be followed over time? What processing has been done to produce comparability over time at object level?	6.4
6f Objects' variable values comparable over time	Longitudinal quality	Can changes at object level be described by the variable?	6.4

10.4.9   Quality indicators – conclusions

Often the persons with the main responsibility for the register have acquired qualitative knowledge about the quality of the register. This knowledge has been acquired through their subject matter knowledge, observations and experience. This means that *they are familiar with the occurrence of a number of types of error without being able to measure their frequency or degree of importance.*

This type of knowledge should also be documented, both to warn users against uncritical use and to indicate where improvement is needed. Lindström (1999) proposes the following levels for the gradation of knowledge concerning different aspects of the register:

Chart 10.10 Levels of knowledge about quality

1. No knowledge at all
2. Vague qualitative information based on appraisal
3. Documentation of the production processes and their characteristics
4. Systematic qualitative information based on surveys
5. Quantitative indicators, e.g. extent of mismatch
6. Quantitative measures of the quality of the estimation

The groups of indicators presented in the charts on the previous pages can be used to describe the characteristics of the register on any of the levels 2–6.

What do these indicators tell us about the quality of the register?

The 40 indicators in groups 1–6 offer a useful reference when deciding how to investigate and describe the characteristics of a particular register. Since the conditions (i.e. the administrative sources and the possibilities of using the register system) differ from one register to another, the important indicators will not be the same for different registers.

Our intention is not that all indicators should be used for all registers, but that the indicators which are important to the particular register are chosen. This limited number of indicators is then used to describe the quality of the register in question.

Many of the quality indicators given above are unique to registers and register-based surveys and are not applicable to surveys with their own data collection.

10.5   OVERALL APPRAISAL OF QUALITY

Register-based statistics can sometimes be used through the calculation of one summarising value:

Employment in the manufacturing industry in our municipality has increased by 1.6%.

But often a number of large tables with hundreds of estimates are studied in an effort to discover patterns of relationship and of time series patterns. The patterns found give rise to conclusions:

The labour structure in our municipality has become worse with respect to age and level of education, hampering the high-tech sector.

Such a conclusion is based on many comparisons with other municipalities and over time. Here it is impossible to conduct a discussion in terms of numerical errors in all of the hundreds of estimates on which the conclusion is based. With every error being divisible into many components each of which is to be judged on its own, the situation appears more daunting than ever. What the user wants to know is whether the conclusion in question can be drawn or not. Are there any statistical pitfalls that the user has not thought of?

On the basis of detailed knowledge of the characteristics of a register, an overall appraisal can be made of the quality of the register and the quality of the register-based survey.

Quality of registers and register-based surveys
The quality of the *register* should be described in general terms, so that potential users can see whether it suits their purposes. The description should relate to the various areas of application that may be of interest. We distinguish between three ways of using registers and the corresponding quality aspects:

– *Cross-sectional quality:* what comparisons can be made within the register?

– *Time series quality:* what comparisons can be made over time on the aggregated level?

– *Longitudinal quality:* what comparisons can be made at micro level over time?

The quality of a *register-based survey* should be described for *one* particular use of the register. Is the quality of the estimates good or bad for this intended use? The relevance and accuracy of the estimates should be described. The following chart compares the ways of describing overall quality:

Chart 10.11 The overall quality of registers and register-based surveys

Quality	Register	Particular register-based survey
Relevance	Only definitions are given	Are the definitions adequate and functional? This is discussed in detail
Cross-sectional quality	What comparisons can be made?	The quality is described only for the particular use.
Time series quality	What comparisons can be made?	Is the quality good or bad? The
Longitudinal quality	What comparisons can be made?	quality of estimates are described

10.5.1   An overall appraisal of the quality of a register

An overall appraisal of the quality of a statistical register should contain the following:

– Definitions of the register population and variables used must be available and easy to understand.
– What comparisons can be made within the register?
  Example: If, in the area of regional statistics, there is a desire to compare municipalities, is the item nonresponse roughly the same in all municipalities or is it possible to make comparisons using imputed values?
– What comparisons can be made over time on an aggregated level?
  Example: Is the item nonresponse roughly the same for all years or can comparisons be made over time using imputed values? Have there been changes in administrative data that make comparisons over time more difficult? Has the register processing been carried out in order to increase the comparability between years?
– What comparisons can be made on micro level over time? Correct longitudinal comparisons place the highest demands on a register. What checking and processing has been done in order to make this possible?

Such an overall appraisal calls for thorough knowledge of the register, which means that detailed documentation is necessary.

10.5.2   An overall appraisal of the quality of a register-based survey

An overall appraisal of the quality of a register-based survey should contain the answers to the following questions:

– How is the survey's *target population* defined? Is the definition adequate in view of the purpose of the survey? Are there any important differences between the target population and the register population?
– How are the survey's *variables* defined? Are the definitions adequate in view of the purpose of the survey? This discussion is directed towards the variables that are of greatest importance for the survey, the most important aggregating and spanning variables on which the analysis is based.
– What *comparisons* are to be made in the survey? Is the register of sufficient quality for these comparisons?
– Are the *estimates* made in the best way? For the sake of both cross-sectional and time series quality, it is important that multi-valued variables and missing values are dealt with using appropriate methods of estimation. In the case of time series quality, it is also important that level shifts in time series are linked.
– Are the *results* interpreted reasonably in the light of the uncertainty of the estimates? Tables based on statistical registers are not to be regarded as offering exact information – there can be random variation and other sources of error, so it is important to be careful of not over-interpreting the results.

Such an overall appraisal calls for knowledge of the quality of the register and for thorough subject matter knowledge. This means that the person carrying out the appraisal must be well acquainted with the research objectives in question.

10.6   MAIN QUALITY ISSUES IN DIFFERENT KINDS OF SURVEYS

As we mention in Section 10.2, how well the data collection process works, primarily determines the quality of a sample survey or census. This means that sampling errors, measurement errors and nonresponse errors are important quality issues here. In Sections 5.4.8 and 5.4.9 coverage errors are discussed. As frame populations generally are based on early available but less reliable sources, coverage errors are a more serious problems in surveys based on data collection than in register-based surveys.

In register-based surveys administrative data and registers are used for statistical purposes. Administrative registers are processed so that objects sets, object types and variables meet statistical needs. The definitions of register population, objects and variables in a statistical register determine the relevance errors of the register-based surveys that use the register. These relevance errors can be judged from the documentation or the register as describes by the quality indicators belonging to group 6 in Section 10.4.8.

Many different sources are integrated when statistical registers are created. The quality of the linkage variables, the sources and the methods used, determine the integration errors in the new register. Integration errors are described by the quality indicators in group 3.1, 3.2, 4 and 5 in Sections 1.4.5 to 14.7.

The discussion above is summarised in the chart below. In our opinion, the main quality issues for register-based statistics are relevance errors and integration errors.

Chart 10.12 Main quality issues in different kinds of surveys

Sample survey	Census	Register-based survey
Coverage errors	Coverage errors	Relevance errors
Nonresponse errors	Nonresponse errors	Integration errors
Measurement errors	Measurement errors	
Sampling errors		

CHAPTER 11

Metadata and IT-systems

All surveys, data matrices and databases need to be documented. This documentation work creates *statistical metadata*, or information describing the statistical data and the survey process. We distinguish between *micro metadata*, which describes the content in data matrices with microdata (i.e. data referring to individual statistical units or objects), and *macro metadata*, which describes the content in statistical tables (i.e. data referring to macrodata that has been formed by aggregating data for groups of objects). Here we only discuss micro metadata.

Micro metadata is needed both by those working with a survey and by those who are users of the survey. However, we only discuss the metadata needs of those working to create statistical registers. We discuss the register system's need for metadata, rather than the technical solutions.

After discussing metadata, we look at IT-systems in general. Until now, we have been discussing statistical registers on a conceptual level, i.e. how registers and data matrices should look and work in a logical sense. In this chapter we continue this conceptual reasoning but we also discuss physical realisation with the creation of databases and applications.

11.1 PRIMARY REGISTERS – THE NEED FOR METADATA

Statistical registers are created by integrating different source registers. Register-based surveys place special demands on metadata, which differ from the metadata needs of surveys with their own data collection. The example in Chart 11.1 below shows the sources needed to create a register and the need for different kinds of metadata.

Chart 11.1 Statistics Sweden's Income & Taxation Register – the need for metadata

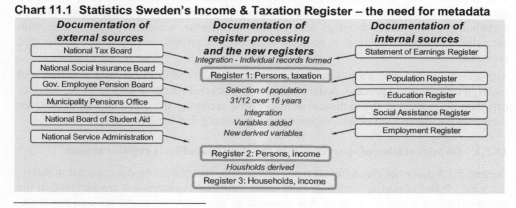

To create the Swedish Income and Taxation Register (I&T), administrative data from six different authorities are used, together with data imported from five different Statistics Sweden registers. Microdata consist of around 500 variables and, to be able to understand these, it is necessary to be well-informed on the tax-related rules that decide the variables' content. New variables can be added as the tax system is constantly changing, and variable names in the administrative sources can change.

This shows that there are significant differences between register-based surveys and surveys with their own data collection, when talking about the nature of the metadata.

It is also necessary to distinguish between the documentation of *registers* and the documentation of *register-based surveys*. When using existing registers to create new registers, *register* documentation is crucial. This type of documentation is characterised by:

- the volume of the metadata which can be very high,
- every administrative source must be documented,
- changes in the administrative system must also be documented,
- the variables can be complicated so documentation must be precise, and
- a large amount of register processing is done to create object sets, objects and variables and this processing should also be documented.

This means that the metadata system must be adapted to suit the requirements of the register system and register-based surveys.

11.1.1 Documentation of administrative sources

Suppliers of the data submit *record descriptions*, etc., which indicate the structure and content of the data being delivered. Furthermore, the receiver should get hold of the *questionnaires with instructions,* which have been used for the administrative data collection. These questionnaires and instructions should be transferred into electronic format. They can then be stored in the metadata system so that everyone who is working with the register can easily have access to them.

Those responsible for contacts with the data supplier should also *interview* them to gain further background information. These interviews should also be documented and stored in the metadata system.

It is important that all changes are carefully noted, and that these are stored over time so that it is easy to gain an overview of the data, so as to judge the comparability over time. A metadata system should therefore also contain a *calendar,* which is an IT system with formalised metadata, where it is possible to search for information by *time, register* and *variable*.

The administrative data should be received, restructured within the statistical office, and undergo a first editing process. A data matrix with the administrative data from the supplier can then be created at the statistical office. Those receiving the delivery should produce their own documentation of this, including the processing that has been carried out.

11.1.2 Documentation of sources within the statistical office's register system

Section 3.3.5 describes the different types of variables that should be documented in different ways. When importing variables from other statistical registers that are included in the statistical office's register system, it is essential to have easy access to the existing metadata, so as to be able to search for and select suitable variables. At the same time as the microdata are imported, the metadata for the imported variables could be easily transferred to the new register's documentation. This would prevent any duplicate work, as the existing documentation can be used again. To be conveniently used again, this documentation must be strictly formalised according to common rules, and stored in a database which is easily accessible.

11.1.3 Documentation of the new register

The new register that is created is documented using the documentation of the administrative sources mentioned in Section 11.1.1, and by importing parts of the documentation from the various registers, from the statistical office's register system, that are needed. The documentation can then be supplemented by descriptions of the processing and integration work that has been carried out and the derived variables that have been created.

Chart 11.2 What should be documented when a new register is created?

All administrative sources

Imports from statistical registers	Register's data matrix/matrices with objects and variables	Register processing

Quality indicators

11.2   CHANGES OVER TIME – THE NEED FOR METADATA

Four types of events can affect register-based statistics and, to avoid incorrect interpretations of time series from register-based surveys, we need to know the following:

- Have changes taken place in the administrative system that make up the sources, so that administrative concepts have been given new definitions?
- Have changes taken place in the way the statistical office has formed the register? Is there, for example, access to new sources, or have new estimation methods been introduced? Several quality indicators should be followed over time so that the data is not misinterpreted.
- Have there been changes to the classifications that are used in the register? For example, the industrial classification or educational codes may have been changed.
- Have any external changes taken place that could have affected the statistics indirectly? If, for example, the value added tax (VAT) rate changes, not only is VAT data affected, but also any data related to private consumption. To be able to interpret the statistics correctly, such changes must be known.

An events calendar should therefore exist, as events that could affect different registers should be documented and compiled in one place. This would make things easier both for those working to create statistical registers and for the users of the register-based surveys, and the risk for misinterpretations of the data would thereby be reduced.

Such a calendar can contain brief details with basic information on:

- what has taken place,
- when it took place,
- which register or registers and which variables have been affected,
- what effect the event has had,
- references or sources for further information.

Chart 11.3 Example of content in an events calendar

Event:	the Adult Education Initiative is introduced
Time:	1997
Register:	Municipal adult education, year 1997
Variable:	New variables added (shown, one item per variable)
Effect:	Increase of number of students by roughly 40%
Event:	New routines give faster information on emigrant persons who have ceased with enterprise activities
Time:	1997
Register:	Business Register
Variable:	-
Effect:	Increased reliability by reduced overcoverage
Event:	Introduction of new Swedish Industrial Classification, SNI 92
Time:	1995
Register:	Employment Register 1993 (corresponding items for all registers concerned)
Variable:	Industry
Effect:	The new SNI 92 replaces the old Swedish Industrial Classification, SNI 69
Event:	Czechoslovakia is broken up into two new states: the Czech Republic and Slovakia
Time:	01-01-1993
Register:	Population Register
Variable:	Citizenship
Effect:	-

The information about *Register* and *Variable* in the events calendar can be used as links to information in other databases with formalised metadata.

11.3   INTEGRATED REGISTERS – THE NEED FOR METADATA

If creating, for example, a longitudinal integrated register with *ten* annual versions of three different statistical registers, it would be necessary to utilise existing metadata effectively. Suppose that the *three* source registers for each year together contain *200* variables, and the metadata system contains *30* definitions for the different register populations and *2 000* variable definitions. This illustrates that register processing can involve large amounts of metadata.

If, for example, registers 1 and 2 have definitions of the register population that has not changed during the ten years and register 3 has a register population that has changed definition once, only 4 (= 1 + 1 + 2) of the 30 possible population definitions are needed.

If, for example, a total of 50 variables have been imported every year from the three registers to the new integrated register, but only four variables have changed in definition once each during the ten year period, then only 54 (= 50 + 4) of the 2 000 variable definitions are needed.

The above example shows the need for an efficient metadata system without large amounts of redundant metadata. The four population definitions and 54 variable definitions needed in this case should be easily accessible.

11.4   CLASSIFICATION AND DEFINITIONS DATABASE

For register statistics, which are based on administrative data, it is especially important to be able to study variable definitions and compare these over time. The I&T register in Chart 11.1 above illustrates the need for this, where around 500 variables based on administrative rules must be managed. Many of these rules and variables change every year.

11.4.1   Classification database

Industrial classification, product category, education, occupation and regional codes are examples of important statistical *standards* and *classifications*. The administrative sources contain data on these hierarchically sorted classifications, and this information is used to create variables within the register system. These classifications are changed at regular intervals. As value sets (sets of all codes or categories) are also large, a classification database is needed to manage all the codes and keys between the different versions. This classification database is an important resource when the variables in a register are documented.

11.4.2   Definitions database and derived variables

In the same way as IT tools are necessary to manage the definition of the statistical classifications; a tool with formalised metadata is also needed to manage the large amount of complicated variable definitions that change over time. We illustrate this with a fictitious example in which three years of an income register have been documented, and also a longitudinal income register (LongI&T), where the three years have been integrated.

Chart 11.4 Documentation of register variables using a definitions database

Register	Variable name	Definition code				
I&T 2001	Sickness benefit	SB1	**Definitions database**			
	Pregnancy benefit	PB1				
	Sick leave pay	SICK1	**Code**	**Definition**	**Definition used**	
					First time	Last time
I&T 2002	Sickness benefit	SB2	SB1	SB1 = "........."	2001	2001
	Pregnancy benefit	PB1	SB2	SB2 = "........."	2002	-
	Sick leave pay	SICK2	PB1	PB1 = "........."	2001	2002
I&T 2003	Sickness benefit	SB2	PB2	PB2 = "........."	2003	-
	Pregnancy benefit	PB2	SICK1	SICK1 = SB1 + PB1	2001	2001
	Sick leave pay	SICK3	SICK2	SICK2 = SB2 + PB1	2002	2002
LongI&T	Sick leave pay 2001	SICK1	SICK3	SICK3 = SB2 + PB2	2003	-
	Sick leave pay 2002	SICK2				
	Sick leave pay 2003	SICK3				

The fictitious income register contains three income variables:

– *Sickness benefit,* where new rules were introduced in 2002,

– *Pregnancy benefit,* where new rules were introduced in 2003, and

– *Sick leave pay,* a derived variable, the total of sick and pregnancy benefits.

The longitudinal register only contains sick leave pay for each year. The chart above shows that variables with the same name, such as *Sickness benefit*, can have different definitions.

Furthermore, variables with different names, such as *Sick leave pay* in I&T 2001 and *Sick leave pay 2001* in LongI&T can have the same definition.

However, because the definition codes are unique (i.e. a specific code is used within the whole register system for one and only one variable definition), there should be no misunderstanding. It is also easier to follow definition changes with a definitions database.

11.5   THE NEED FOR METADATA FOR REGISTERS

Those creating statistical registers within the register system need different types of metadata and practical IT tools to register and use this metadata in their work. The chart below shows nine types of metadata and the tools that could be used.

Chart 11.5 Different types of metadata and tools in register documentation

1. Classification and definitions databases Formalised metadata	2. All administrative sources Formalised metadata Questionnaires, instructions, interviews, etc.	3. Events calendar Formalised metadata
4. Imports from statistical registers Formalised metadata	5. Register's data matrix/matrices with objects and variables Formalised metadata	6. Register processing SQL script with comments
7. Bulletin board An Office system	8. Quality indicators Text documents	9. Documentation system Manages documents

There should be a system that integrates all the formalised metadata which currently exists, but also that which exists in what we mention above: The Calendar and Classification & Definitions databases. In addition, a system is needed to manage documents with other metadata. Systems with *formalised metadata* can be used for the following (the numbers refer to Chart 11.5 above):

1.   Classification and definitions databases with easy access.

2.   Documentation of data matrices from administrative sources.

3.   An events calendar, easy access to information of important changes.

4.   Imports from statistical registers, formalised metadata are easily imported.

5.   Documentation of the data matrices in the statistical register. Register population, object type and variables are described.

The other documentation can consist of *different types of documents* (the numbers refer to Chart 11.5 above):

2.   Text information on the administrative systems, administrative questionnaires with instructions and minutes and notes from meetings with those delivering the registers.

6.   SQL script with comments describing how the register processing is done.

7.   A bulletin board for all those using the registers who find inconsistencies and errors. All those who support the base registers, according to Section 5.4.7, should add their contributions to the respective base register's bulletin board.

8.   Quality indicators, the most important indicators for the register in question.

9.   All documents above are managed by a special system for easy access.

Uniform text documents

Data matrices created via collecting data in sample surveys are usually documented in text documents structured in a uniform way. The chart below illustrates how this kind of documentation can be structured to suit register-based surveys. The chart compares the most important part of each kind of survey – the data collection process for sample surveys and the integration process for register-based surveys.

Chart 11.6 Metadata for sample surveys and register-based surveys

Sample survey: The data matrix The data collection process	Register-based survey: The register The integration process
1 Frame and frame procedure 2 Sampling procedure 3 Questionnaire 4 Data collection procedure 5 Data preparation	1 Describing sources 2 Receiving and editing each source 3 Integration 1 – register population 4 Integration 2 – objects 5 Integration 3 – variables 6 Consistency editing

This chart illustrates how microdata have been created. As microdata are created in different ways in these two kinds of survey, this part of the documentation should differ. The other parts of the documentation can have the same structure.

The metadata system – a survey with data collection

A statistical office collects metadata via special systems from its staff. Those who are responsible of documentation fill in electronic forms, and the result should be metadata of good quality. Good quality means that the metadata system has good coverage, low nonresponse and small measurement errors, and that that the metadata are easy to access and understand.

Defining object types can be a difficult part of the documentation, here measurement errors or misclassifications can arise if concepts are difficult to understand. The distinction between *object type* and *variable* should be made clear to those who report metadata; otherwise the users of metadata will get problems when they are searching data about one particular object type. If an object type has been defined as another kind of object type or as a variable, the user will not find the desired metadata. To avoid such misclassifications or measurement errors, there should be only a few object types defined in the system, and these should be easy to grasp.

Example: A school child can be defined as a relational object *person • school* or as a *person*. We suggest that schoolchildren are defined as persons, and then it is easy to find other registers on persons with more variables concerning these schoolchildren.

Example: Products can be defined as *variables* connected with enterprises, or as an *object type* product. We suggest that products are defined as variables – enterprises produce products of different kinds; the value and quantity produced of each product can be defined as enterprise variables, and these can be combined with other enterprise variables from other registers.

In an IT environment, the term 'object' is used frequently, and sometimes with definitions that differ from the statistical term. In a database solution, rows in certain database tables are called 'objects' without being objects in the statistical or conceptual meaning. This can cause misunderstandings. When a survey is documented, only objects that are part of the register population should be called objects in the statistical part of the documentation.

11.6   IT SYSTEMS FOR REGISTER-BASED STATISTICS

When developing an IT system to support the creation of statistical registers the register documentation is crucial. It is therefore essential that documentation work is underway before the systems development work begins. The various types of formalised metadata mentioned above are then available, and can be utilised using IT tools available for managing metadata.

In this section analysis, design and physical realisation of IT-systems are discussed.

11.6.1   Analysis

The analysis process aims at creating a basis for the design and realisation of the system by analysing and formalising the demands that it needs to fulfil. The method used is modelling – identifying, discussing and documenting – the system's objects and processes.

Object modelling identifies the input and output data used in the system, and describes the connection between them.

Process modelling illustrates the system processes, and describes their sequence and the dependencies between them. This stage describes how the data is to be collected. It also documents how users should, or would like to work, by describing *use cases* in a standardised way.

An IT system for register-based statistics differs from one that supports a survey with its own data collection, primarily with regards to the modelling of the data collection process. Instead of deciding on a frame population and selecting a sample, in this case the process involves determining how well the registers that make up the main data sources cover the need for output data. If further data is needed, additional sources must be identified and documented. The main task at this stage is modelling of the selection and matching processes needed to create the register.

This phase should also investigate how the cooperation with other systems is to work. The primary cooperation that must be explained is within the register system: what dependencies exist to and from other registers, how to manage conflicts between information sources, etc. Additionally, the register's role as a source for other registers within the register system must be explained.

Handling the *time dimension* (see Section 3.2.4) is a complication that is typical for statistical registers. A regularly recurring survey with its own data collection will (put simply) create either a new file or database table per survey round, or add new entries to an existing file or database table. A statistical register, on the other hand can, if the underlying registers permit, change continuously by adding, deleting or altering data. The analysis process must clearly define what reference time information will be necessary or expected for regular production, as well as anticipated ad hoc tasks. This will provide a basis for decision on which types of information are needed in this register: the current population, populations at certain and arbitrary points in time, calendar year population, events and longitudinal information.

The analysis process should also show the level of detail needed for information used in different parts of the system. The possible, desirable and meaningful frequency to use for updating the statistical register from the underlying register should also be documented here.

We want to emphasize the importance of maintaining an overall view of the register system during the analysis phase. Even if the development work is focused on one specific register, it is necessary to constantly look around and take note of the requirements and needs of the total register system.

11.6.2 Design

Based on the demands on the system, the design process aims at deciding on:

- the choice of technology for the total systems solution,
- any divisions into sub-systems,
- the database design, and
- an architecture for every application or processing.

Choice of technology

Defining the cooperation between a statistical register and its source registers is an important part of the system design. When designing a primary register, the technical cooperation with the administrative register is particularly important. This is dependent on several factors, such as:

- The storage technology used by the administrative register: Can data be transferred directly or must they be converted?
- The updating frequency: If data is only to be transferred on rare occasions, maybe once a year, a relatively simple manual solution can be used. However, a higher frequency of updating will require more automation and accuracy from both the sender and the receiver.
- Data quantities: The transfer of a large amount of data would naturally demand more from the technical solution. A small amount can be transferred by email, CD-R or such like, while larger quantities require well-planned special solutions.
- Security/unpermitted access: All microdata relating to individual objects are strictly confidential, and must be handled according to current confidentiality regulations.
- Security/vulnerability: This refers to the risks of losing data when transferring using, for example, inadequate routines or technology, and also the risks associated with loss of competence on both sides.
- Communication technology: There are often a number of possible technologies available on any occasion. When choosing the technology, the previously mentioned factors should be taken into consideration.

Apart from the register being able to cooperate with one or several source registers, the technical aspects must also work with other registers in the register system. The technology and solutions that have previously been used in closely related registers will therefore have a great influence on the solutions for the register now in question.

The analysis and combined evaluation of these demands, desires and possibilities can cause considerable differences to the need for work input during the next development stages.

Database design

The design of a system's databases is crucial for how well the system will work. Thus, it should always be carried out by systems developers with wide experience in data modelling.

Database design for an IT system for register-based statistics does not differ fundamentally from database design for other IT systems for statistical production. The system should be optimised for:

- ease of use, user-friendliness,
- reliability and performance for uploading data, new and updated values,
- processing, i.e. editing, derivation of variables, etc., and
- output, i.e. production of statistical tables, other standard or ad hoc tasks.

This often places conflicting demands on the database design, and often results in dividing the system into several databases. In simple terms, we often talk about three main types of databases optimised for different uses: input, processing and output databases.

- An *input database* should be able to receive all types of updates, including completely incorrect input data. Input databases therefore normally have only a minimum number of consistency rules. Data is checked and corrected if necessary before it is sent on.
- The *processing database* is often normalised, i.e. it has strict consistency rules and contains no duplicate data. This is a model which is optimised for editing micro data with minimal risk for introducing inconsistent data.
- An *output database* should be optimised for the most commonly occurring types of selections, analyses and other processing. Once data has been loaded, no changes should be made. Thus, in terms of content, the output database is synonymous with the completed data matrix. Duplicates are permitted if they improve performance and/or facilitate things for the users.

The data format for all variables should be determined in this phase. In most cases, this does not cause any problems, but for linkage variables and particularly those that can be used for linking between different registers, it is very important to follow established standards, i.e. the Swedish personal identification numbers must be stored using exactly 12 digits.

The time dimension, which is discussed in Section 11.6.1 above, can greatly affect the database design. If the system has high demands for time information for every amendment and covers a long period, there is a risk that using one single output database to carry out all types of output processing can result in a user unfriendly system with a poor performance level. In such cases, the output database may need to be divided up even further, for example as follows:

- a total or historical database,
- one or several databases for specific points in time, such as the end of the year and a database for the calendar year version of the register, and
- a database for the current stock version of the register.

In the same way, comprehensive commissioned work based on the register can result in the need to create separate project databases. Carrying out ad hoc tasks and routine processing in the same database may cause serious disturbances, which should be avoided.

Chart 11.7 Different types of databases for a statistical register

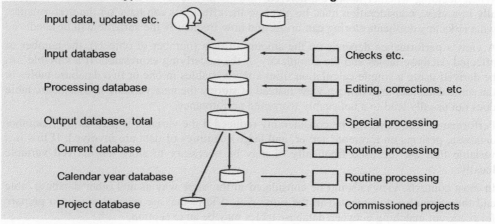

Input data, updates etc.

Input database — Checks etc.

Processing database — Editing, corrections, etc

Output database, total — Special processing

Current database — Routine processing

Calendar year database — Routine processing

Project database — Commissioned projects

The requirement for cooperation within the register system obviously sets certain limits on the choice of database management system. Software that, in certain respects, has excellent performance or that works well for a particular register in some way may still not be chosen, if it works badly – or not at all – with the software used by closely-related registers in the register system.

Modelling for the register system – not the register

Throughout this book, we have emphasised the importance of looking at the whole picture, the register system, instead of just at the individual register. We can go further than this to state that the register system is a key part of the *Statistical Data Warehouse*.

The term *Data Warehouse* has primarily been used in the business world to describe a coordinated, consistent and quality assured data store, often relating to longitudinal analyses of mainly financial data.

The Data Warehouse typically contains large quantities of data, divided into a large number of information categories. When designing one or several databases for a Data Warehouse, it is preferable to minimise the need for matching more than two database tables at the same time, partly because matching is logically and syntactically hard for users, and partly because it deteriorates the performance of the database management system.

Derived variables and views

As long as the system design is conceptual, or logical, derived variables are handled in the same way as other variables. An important part of the physical database design is to determine which derived variables should be stored permanently, and which should be combined and calculated on demand. The latter alternative can be completely ad hoc or can be prepared for better user-friendliness by creating *views*. A view is a 'virtual database table', i.e. a pre-defined expression that appears to the user like any database table but that, in practice, carries out operations on one or several underlying database tables.

Example: Database table X contains data on a variety of income types, working hours, etc. From the data in X, we can derive the variable *full-time salary*, which we place in view Y. Users can then use a simple query to select the information on full-time salary from Y, although every time the calculation will be carried out in X. The advantage with a view is that it is easy to understand and simple to use. A disadvantage can be that it may require a high level of machine performance to work well.

When deciding whether to store a derived variable physically in a database table or virtually in a view, consideration must be given to the effect this can have on the performance, what risks any duplicate storing can bring, and how frequently the variable will be used.

A view's performance depends on the amount of data (number of objects), the number of affected database tables and the complexity of the underlying expression. If a variable can be derived using a simple calculation from a few variables in one or two database tables in the same database, then using a view instead of storing the new variable in a database table does not usually lead to a noticeably worsened performance.

Performance can, however, be considerably reduced if the variable is derived from another database, perhaps on a remote server, and large quantities of data are involved. If this is a variable that will be used frequently, it may be necessary to store the derived variable locally.

In most contexts, views should be considered in the same way as any other database table and therefore also documented in the same way. Views that are created purely to prepare for recurrent matching between database tables may be an exception.

Application architecture

During this phase, the need for the development of tailor-made applications should also be investigated. A register with very frequent updating of input data would need an automated process for transfer and loading, while a register with irregular or, for example, annual updating can be done with manual routines. In the same way, standardised, often recurring routine processing can be simplified with the development of an application that can be managed by 'anyone' instead of needing to repeat similar manipulation sequences for every task.

11.6.3   Physical realisation

The physical realisation primarily involves programming and establishing the necessary databases to prepare the system for use. Programming as well as creation of databases should be carried out with recommended tools following established standards.

Physical optimisation of the register system

The databases that constitute the register system should be modelled, to logically cooperate well, and must also physically work well together. It is not possible to implement the register system in one physical database, probably not even on one server. The physical cooperation will always be affected by whether matching is taking place on one server or between different servers.

The tools for allowing users to interpret a group of servers as one logical unit are being improved all the time, but the fact remains that a server works many times faster internally than any communication link between servers can. A judgement must be made, on which registers in the register system have the greatest need to work together, and therefore should be localised on the same server.

In the example in Chart 11.8, information in registers R1–R4 is regularly used together, as is the information in registers R5–R8, and therefore the databases are grouped accordingly on the two servers.

Chart 11.8 Registers that are used together are placed on the same servers

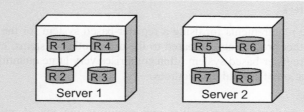

A location of registers as shown in the example does not make it impossible to use registers R4 and R5 together, for example. It will be slightly more complex and can take more time. One method for retaining the usage within the two servers in the example, and also improving the usage between the registers R4 and R5, is to keep copies of these registers on the other server.

A controlled double storage in this way can be managed in several ways. The safest is probably to use the database management system's built in replication technology. In the example in Chart 11.9, every change in register R4 in Server 1 will also be carried out in register R4 in Server 2 at a specified point in time, such as every hour, every night or every week. The same applies for register R5 in Server 2 and its replicated copy in Server 1.

Chart 11.9 Duplicate storing of some registers that are used together

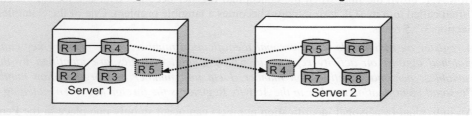

Indexing

With all processing of data from a data store based on relational databases, a correct and appropriate *indexing* is vital for performance. Without appropriate indexes, the database management system will, when matching between two database tables, for example, always evaluate every possible combination before making a choice – even if the user only wishes to match one row in every table. Indexing the primary and foreign keys, however, will provide the system with necessary information to directly locate the two matching items.

The database tables are normally automatically indexed on primary and foreign keys by the data modelling tools. In a processing database, these indexes are often the only ones needed to create a good environment for editing data. Further indexes on other variables may even worsen the performance.

However, in an output database, where changes are uncommon and well planned, and where searches and matching may have to be carried out on variables other than the keys, well planned indexing may return significant performance improvements. This is achieved at the cost of worsened performance when updating.

11.6.4 Use of registers

There are no specific requirements for using a register-based system for the production of statistical tables or other processing compared to the use of other systems, except possibly in two respects. The register-based system often comprises very large quantities of data and often involves integration of several data sources.

Large data quantities

Processing or retrieving large data quantities may require some considerations. The time required for processing often has a nonlinear dependency on the size of the data quantity. The time often depends on the hardware, the software, the configuration and the programming method. Limitations in memory capacity can in difficult cases, for example, result in an exponential relation between data quantity and processing time. However, the most important factor for good performance is often the user's way of working. With data quantities of a modest size, the user can often use trial and error with the help of test runs, which is not possible if the data quantities are large.

Matching

Matching information from several registers or between a register and a sample survey is a common task. On a conceptual level, it looks fairly easy, provided that the necessary linkage variables are available.

However, down the line, when the conceptual register becomes database tables, matching (often called *join* in SQL terminology) becomes a tangible problem. One of the examples in Section 3.3.5 reads:

A register on individuals with personal identification number as the primary key can be matched against an activity register in which gainful activities are identified by three variables: personal identification number, local unit number and organisation number. Personal identification number in the Activity Register is the foreign key when matching.

An individual (a personal identification number) can occur in only one place in the Person Register but in one or several places in the Activity Register. We therefore have a one-to-many relationship between objects. The situation can be described graphically, as follows:

Chart 11.10 Matching two registers

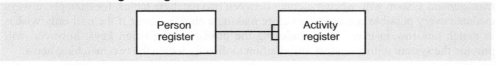

When this is physically implemented, we can first assume that the data we want to collect exists in only one database table in Person Register (called *Person*) and in only one table in the Activity Register (*Activity*). Matching the two database tables still seems fairly easy:

Chart 11.11 Relationship Person – Activity

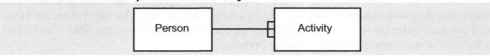

Matching can be expressed in SQL language as a join. Several end-user tools, such as SAS and SPSS, support uncomplicated joins without the user being required to write SQL code, provided that the keys were specified correctly when creating the database tables. In most cases, these tools are preferable to writing SQL code, as the support provided by the tools reduces the risks for missed links, other mistakes and typing errors.

These database tables, however, usually only contain the numerical codes of qualitative variables, but in many cases we want to include the actual explanatory text that is stored in the metadata tables. Immediately, the chart in our simple example becomes more complicated, as it requires a join between four database tables:

Chart 11.12 Matching two registers with metadata

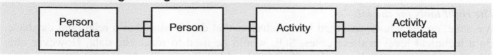

In reality, this would probably be further complicated by, for example, the storage of the activity variables we are interested in, being spread among several database tables, or some variables being aggregated while being matched, etc.

A further common complication is the handling of *mismatch* (Section 3.3.5). In our example, it could be that objects in the Activity Register are missing for some reason, or have an incorrect personal identification number, and it is almost certain that some objects in the Person Register lack a corresponding object in the Activity Register.

The standard situation in SQL, as with all the end-user tools (SAS, SPSS, etc.) is that only complete matches are selected. If we want to select all rows from the database table *Person* and add data from *Activity* when they exist, but otherwise leave empty (*null value*), this must be specifically indicated using the SQL function *outer join*. Syntactically – and logically – the use of outer join is limited, i.e. in combination with aggregation and with complete matches (*inner join*).

Chart 11.13 Inner and outer join

Inner join – only select rows where nr in Person matches nr in Activity:	*Outer join* – select all rows in Person, add matching information from Activity if such exists:

select Age, Salary
  from Person
    join Activity
      on Person.nr=Activity.nr
	select Age, Salary
	  from Person
	    left outer join Activity
	      on Person.nr=Activity.nr

Our simple example here already shows that the technical, or syntactic, level of difficulty increases quickly. This can sometimes make it difficult to determine if the result produced by advanced matching is actually correct. A carelessly and incorrectly written SQL code can, at worst, give a result that on the surface appears correct – the response has the correct number of rows and values in the correct order – but is still incorrect. It is therefore nearly always preferable to carry out a task using several simple, controllable steps, storing intermediate results temporarily and checking them between every stage, instead of carrying out everything in one giant advanced step. It is more important to ensure that the result will be correct rather than to write fancy code.

Another method for simplifying complicated processing is to prepare common matches by creating views. Using a view, a complex SQL expression can be hidden so that the user sees a simple database table containing all the variables that are of interest. In our example, we can once and for all create the view *PersonActivity*, which contains our join between four tables. Then we simply refer to the view instead of the four tables every time we need to access the data. The data matrices for a 'combination object', discussed in Section 9.2

can beneficially be implemented as a view, which can be used for the retrieval of weighted values.

Apart from the syntactic problems of actually achieving what we intend to do, the content-related problem of selecting the right rows from the database tables often occurs when matching. In our example, we can assume that the Activity Register contains a time dimension. A syntactically correct matching would still give a nonsense result if we forget to indicate which point in time, etc. that is required.

The right tools for the job

We have, on the whole, assumed in this chapter that data is stored in a relational database. On a technical level in these cases, SQL would be used to reach the data. However, this should not be interpreted such that all users should carry out all processing with pure SQL directly in the databases. Instead, it is often appropriate to use a tool that gives a more user-friendly interface (click, choose from list, drag and drop, etc.) to gather and manipulate data, and which gives adequate support for matching.

In addition, it is not always certain that letting the database management system carry out all the work is the most effective method. This system is optimised for working in a *quantity-oriented* way. This is its normal, very effective working method. It is possible to force the database management system to work *sequentially* so that it can manage tasks that require this behaviour, i.e. selecting samples but, with this working method (stepping forward using a *cursor*) the system becomes ineffective. In such cases and, in many cases involving statistical processing, such as estimation with weights, it would be most appropriate to use, for instance, SAS.

Unfortunately, there are many situations in which it is not entirely clear which tool is the most appropriate. The most commonly used tools overlap considerably regarding their areas of application. Many tasks may be carried out in three or four different ways with as many tools. The user's skill and experience is often more important to the result than the tool's objective suitability when carrying out a specific task quickly and securely. The disadvantage is that solutions risk becoming dependent on the individual.

CHAPTER 12

Protection of Privacy and Confidentiality

Population censuses in Sweden until 1990 have been 'traditional' censuses, with question-naires sent to every household. Since the Population and Housing Census in 1970, debates about privacy have arisen in conjunction with every new population census. These debates also increased the nonresponse for other surveys, such as the Labour Force Surveys. This shows that anxiety about privacy can be a threat against the operations of a national statistical office. It can be noted that even in Sweden, where there is a long tradition with public documents at the authorities being accessible by everyone, this anxiety can arise.

Today, there are many new threats to privacy – when you make a phone call, this is registered; when you use your credit card, this is registered; and when you walk in the street in many cities, cameras follow you. These new kinds of registration have no connection with statistical offices, but even if statistical registers are less emergent today, it is important that statistical offices maintain high standards for protecting the privacy of both people and enterprises.

Even if the real risk of threatening people's privacy with statistical use of administrative data is comparatively small, a debate in today's media can have serious consequences regarding the reputation of a statistical office.

We use the term *confidentiality*, which applies both to individuals and enterprises. When the register system is improved to facilitate the integration of sources, with the aim of raising the quality of the statistics, the protection of confidentiality must also be strengthened. This protection can be improved in the following ways:

1. Existence of variables with *text* is minimised.
2. Existence of identification variables such as *official identification numbers* should be minimised.
3. The risk that data regarding individual persons or enterprises can be derived from the *statistical tables* in publications or official databases should be minimised.
4. Before researchers, after appeal, are given the possibility to analyse data matrices with *microdata*, these data matrices should be processed to minimise the risks for the disclosure of information on individuals.

The protection of confidentiality can, according to points (1) and (2) above, be increased in two ways: by minimising the number of databases with text or official identification numbers; and by minimising the number of people with access to such databases.

Register-based Statistics – Administrative Data for Statistical Purposes   A. Wallgren and B. Wallgren
© 2007 John Wiley & Sons, Ltd

Data submitted to statistical offices for the production of statistics are covered by statistical confidentiality in each country's legislation. As well as this legislation, the data processing within a statistical office should be organised so that confidentiality is protected in an efficient way. We discuss here the internal security regarding register data, and the risks of disclosure connected with published tables and micro data used by researchers.

Statistical disclosure control is a vast and important field, which is described, by for example the Federal Committee on Statistical Methodology (2005) and Willenborg and de Waal (2001).

12.1   INTERNAL SECURITY

The number of databases with names, addresses and other information in text should be minimised. The use of official identification numbers, such as social security numbers or social and fiscal numbers, can also be minimised if the statistical office recodes these official numbers into their own record identification numbers. If the number of people that have access to such databases is also minimised, the internal security is organised to protect confidentiality.

12.1.1   No text in output databases!

The administrative data delivered to a statistical office can contain names, addresses and other text data where the administrative authority needs these for their administrative purposes. The salary registers from employers can also contain data in plain language. When the preliminary register processing is carried out, these details should be replaced by codes.

A fictitious example can illustrate this:

Chart 12.1 An administrative register – unprocessed data in the input database

PIN	Name	Address	Post Code	Enterprise, local unit	Job title	TNS code	Actual salary	Extent of work
5602301234	Pson Per	1st Street 7	111 11	Statistics Sweden, Stockholm	IT-specialist	4321	18340	0.60
6706312345	Ason Eva	2nd Street 2	777 77	Statistics Sweden, Örebro	Head of department	1234	45780	1.00

As part of the editing of the data, the personal identification number (PIN) should be given the correct format for the statistical office, all plain language should be replaced by codes, and all variables that are not statistically interesting should be deleted. In addition, statistically important variables are imported and derived variables are formed.

In this example, the Swedish PIN in Chart 12.1 consists of date of birth (yymmdd) plus four individual digits. The correct format within Statistics Sweden also includes the century, as in Chart 12.2. The names and identity numbers of the persons in Chart 12.1 are checked against the Population Register, and thereafter the name is no longer needed, as all matching within Statistics Sweden is done with the PIN-variable.

The home address and address of the local unit in Chart 12.1 are replaced with regional codes in Chart 12.2, and job title and administrative codes (TNS) for occupation are replaced with statistical ISCO-codes for occupation (see Section 6.2.2). A derived variable for full-time salary is also created in Chart 12.2, and educational code is imported from the Education Register.

Chart 12.2 Corresponding statistical register – processed data in the output database

PIN	Residential Municipality	Local unit id number	Local unit Municipality	Occupation ISCO	Education code	Actual salary	Extent of work	Full time salary
195602301234	0180	12345678	0180	2222	1234567	18342	0.60	30570
196706312345	1880	23456789	1880	3333	7654321	45780	1.00	45780

In this example, it would be advisable to have access to the original job title for any future changes in occupational classifications, for example. The simplest method would be to take the administrative sources out of the data system and to archive them in a locked space once the statistical register has been edited and processed. This would make it possible to carry out future controls without the administrative data being available in the data system.

Names, addresses and other details in plain language, relating to individuals or enterprises, should only exist in the input databases. Only a small number of people within the statistical office should have access to these. The output databases contain the statistical registers, and names, text, etc., should as a rule not be allowed there.

Chart 12.3 Different types of databases with different kinds of access

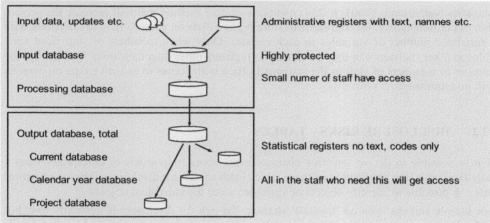

Names, addresses and other details in plain language, relating to individuals or enterprises, should *only* exist in the base registers output databases. Only staff working with interviews and questionnaires should have access to name and address details. *All* output databases of registers apart from the base registers should be completely anonymised by the use of codes.

12.1.2 Existence of identity numbers

In addition to thoroughly checking the people who have the authority to work with microdata, and minimising the existence of variables with text, the use of object identities numbers should also be restrictive. Statistics Denmark and Statistics Netherlands have decided to replace the official identity numbers of persons with record identity numbers, created and used by the statistical office. This is done to prevent misuse of social and fiscal numbers or PINs, which is discussed in Statistics Netherlands (2001, p. 249).

Personal identification number

In the first place, we should reduce the use of personal identification numbers. These numbers exist in the administrative sources that are delivered to the statistical office. They are found in administrative registers on individuals, in some cases on small enterprises, on activities and in the real estate register. The number of people receiving this administrative data at the statistical office should be limited.

The administrative data is transferred, after editing and processing, to the input database, the processing and output databases. In all input databases with personal identity numbers, these numbers should be replaced by the statistical office's RIN, the unique Record Identification Number for each individual. There should be one central database, with restricted access, where this recoding is supported.

In countries where personal identification numbers are not used, name, address, birth date and birthplace are used when matching records in different registers. These identifying variables can be replaced by Record Identification Numbers in the same way as the official PINs are replaced with the aid of a central database.

After this replacement of sensitive identifying variables, all output databases with data regarding persons can be integrated with the RIN. The main part of the staff working with register-based statistics will thus have no access to names, etc. or official identification numbers.

Statistics Netherlands (2001, p. 254) mentions another method that will increase the protection of confidentiality. It is desirable to store register-data on persons in many registers with a restricted number of variables in each register. Only when members of staff need variables in other registers can they get access to registers other than their own. In this way, the number of members of staff at the statistical office with access to sensitive data on persons will be minimised.

12.2 DISCLOSURE RISKS – TABLES

If it is possible to derive sensitive characteristics about individuals or enterprises from a statistical table, the publication of these tables leads to risk for disclosure. Disclosure means that it is possible to identify objects or variable values for individual objects.

For sample surveys with no 'take-all' stratum, the risk for disclosure is generally low, but tables based on censuses or register-based surveys must always be checked so that details about individual objects cannot be disclosed.

12.2.1 Rules for tables with counts, totals and mean values

In the table below, fictional data is shown relating to enterprises in a specific industry in a specific municipality.

Chart 12.4 Enterprises by size category within industry Y in municipality X

Number of employees	Number of enterprises	Number of employees	Turnover SEK 000s	Wage sums SEK 000s	Average turnover per enterprise	Average annual salary per employee
	(1)	(2)	(3)	(4)	(5) = (3) / (1)	(6) = (4) / (2)
0–9	9	50	28 250	11 800	3 139	236
10–99	5	190	116 900	43 380	23 380	228
100–199	3	615	391 650	151 200	130 550	246
200–499	2	600	287 000	169 400	143 500	282
500–	1	705	240 550	159 400	240 550	226
Total	20	2 160	1 064 350	535 180	53 218	248

There are different rules that can be used to decide if the risk of disclosure fore some cells in a table is too high. We illustrate below three rules with data in the table above. If cells have too high a risk of disclosure according to a rule, they should not be published.

The table can either be redesigned with fewer class intervals or some cells should be suppressed. More alternatives also exist; the table can be altered in a number of ways described in the Federal Committee (2005), and Willenborg and de Waal (2001).

Two rows in the chart above disclose data on individual enterprises:

- In the row with only one enterprise, this enterprise can be identified by the variables municipality, industry and size category.

- The row with two enterprises with 200–499 employees. People knowing of one of these two enterprises would be able to identify details about the other using simple subtraction.

The data in columns (2)–(6) of these two rows should therefore be suppressed. Three frequently used rules can be used to determine if a cell is sensitive, which means that the risk of disclosure can be high, are described below. More than one of these rules can be combined into a more complex rule.

1. The threshold rule

A cell in a table can be defined as sensitive if it is based on less than k observations. Such cells should be suppressed, but also other cells that make it possible to calculate the frequencies in the cells that should be suppressed

Example: With $k = 3$ the two bottom lines in Chart 12.4 are sensitive. If it is judged that column (1) will not reveal any sensitive information it can be published, but the other columns should not be published.

2. The (n, k) or dominance rule

A cell in a table can be defined as sensitive if the n largest values sum to at least $100k\%$ of the cell total. This rule is applied to cells with totals or sums.

Example: According to the rule $(1, 0.75)$ the last line in Chart 12.4 is sensitive as the enterprise is 100% of all sums. There are two enterprises with 200–499 employees. If the larger of these two has the attributes shown in Chart 12.5, column (4) in the line with the two enterprises with 200–499 employees is sensitive, as $137\,000/169\,400 = 0.80$, and columns (4) and (6) on this line should not be published.

Column (3) is not sensitive as $195/287 = 0.68$, and column (3) and (5) in Chart 12.4 can be published according to this rule.

Chart 12.5 One line from Chart 12.4

Number of employees	Number of enterprises	Number of employees	Turnover SEK 000s	Wage sums SEK 000s	
	(1)	(2)	(3)	(4)	
200–499	2	600	287 000	169 400	← Can these cells be published?
The largest of these two:		400	195 000	137 000	

3. The p-percent rule

A cell in a table can be defined as sensitive if it is possible to estimate the value for at least one object in the cell with an error smaller than p%. This rule is applied to cells with totals or sums. In the example below, the simplest version of this p-percent rule is illustrated.

Example: If the two last lines in Chart 12.4 are combined into one class interval, with 200 and more employees, we have the micro data in Chart 12.6 below for these three enterprises. We assume that a person who is working at the medium enterprise tries to estimate values regarding the largest enterprise.

Chart 12.6 The two bottom lines in Chart 12.4 are combined into one line

Number of employees	Number of enterprises (1)	Number of employees (2)	Turnover SEK 000s (3)	Wage sums SEK 000s (4)	
200–	3	1305	527 550	328 800	◄— Can these cells be published?
Smallest enterprise		200	92 000	32 400	
Medium enterprise		400	195 000	137 000	
Largest enterprise		705	240 550	159 400	

The p-percent rule with $p = 15\%$ will be used. A person with information about the medium sized enterprise can estimate the values regarding the largest enterprise in the following way:

Number of employees is = 1 305 – 400 (known) – 200 (lower limit) = 705
705/705 = 1.00, error smaller than 15%, do not publish!

Turnover is approximately = 527 550 – 195 000 (known) = 332 550
332 550/240 550 = 1.38, error greater than 15%, OK to publish

Wage sum is approximately = 328 800 – 137 000 (known) = 191 800
191 800/159 400 = 1.20, error greater than 15%, OK to publish

The p-percent rule with $p = 15\%$ indicates that cells (1), (3) and (4) can be published.

12.2.2   The threshold rule – analyse complete tables!

The table below shows a number of persons convicted of crimes in a certain town in a certain year. Some cells in this table with sensitive information contain small numbers.

Chart 12.7 Town Z, number of persons convicted of crimes by age and sex

Age	Women	Men	Both sexes
16–24	0	7	7
25–34	1	7	8
35–44	3	17	20
45–54	3	3	6
55–64	3	5	8
Total	10	39	49

With the information in this table, it is not possible to derive any sensitive information about any individual. It is first together with the number of persons in town Z, by age and sex, that sensitive information can be disclosed.

Using the threshold value rule with $k = 3$, it can be supposed that the shaded cells in the table can be published. But this is an incorrect interpretation of the rule, as it is the *complete* table's cells that must be of at least three observations. The basis of a complete table is the total number of persons in the population.

Chart 12.7 above is actually only a part of a larger, more comprehensive frequency table in which the population is broken down by three variables at the same time: sex, age and criminal charge, measured by the categories convicted/never convicted.

In the complete table, the population is simultaneously tabulated by sex, age and criminal charge. It is in this complete table that the risks for disclosure can be seen.

Chart 12.8 Town Z, population by sex, age and criminal charge

Sex	Age	Convicted	Not convicted	Total population	Risk for disclosure
Women	16–24	0	2	2	0.00
	25–34	1	1	2	0.50
	35–44	3	28	31	0.10
	45–54	3	41	44	0.07
	55–64	3	5	8	0.38
Women total	*16–64*	*10*	*77*	*87*	
Men	16–24	7	0	7	1.00
	25–34	7	1	8	0.88
	35–44	17	65	82	0.21
	45–54	3	65	68	0.04
	55–64	5	3	8	0.63
Men total	*16–64*	*39*	*134*	*173*	
Both sexes	**16–64**	**49**	**211**	**260**	

It is clear from the complete table which parts are unsuitable for publishing. The threshold rule from the incomplete table in Chart 12.7 with at least three as the basis means that the white cells should not be published. When looking at a woman aged 16–24, the conclusion can be drawn that she has not been convicted. This information is perhaps not so sensitive, but for women aged 25–34, the threshold value rule would actually protect against the disclosure of sensitive information.

However, in the complete table in Chart 12.8 there are two additional cells with frequencies smaller than 3. The cells with 0 and 1 not convicted men should be suppressed, but also the cells with 7 and 7 convicted men as the suppressed cells can be calculated with these and the margin for the total population.

This means that the cells for men aged 16–24 years old and men aged 25–34 years old should not be published. Even more cells should be suppressed, as it is possible with the other cells and the known margins to calculate good estimates of the sensitive cells. Rules are not enough; judgement is also necessary. The last column in Chart 12.8 shows the probability that a person described by age category and sex is convicted. These risks can be used to find the sensitive cells where judgement is necessary.

12.2.3 Combining tables can cause disclosure

Detailed tables based on registers are available via a statistical office's homepages. If, for instance, a table for a small region is combined with a table obtained by a commission where a table with data for squares maps from the statistical office's geographical information system (GIS), disclosure can be possible. If the small region consists of one GIS square plus forest plus one small population centre, the following tables can be combined:

Chart 12.9 Table for the region

Occupation	Sex	Number of persons	Average income
Statistician			
	F	8	32 560
	M	7	32 780
...	...	...	...

Chart 12.10 Table for the GIS square

Occupation	Sex	Number of persons	Average income
Statistician			
	F	7	29 870
	M	6	30 120
...	...	...	...

By combining these tables, it is easy to calculate the income of the two statisticians who live in the small population centre not included in the GIS square. That tables can be combined in this way must be considered by the staff delivering GIS data to customers. This means that GIS data must be carefully checked before delivery.

Tables produced by a statistical office can also be combined with tables produced by, for instance, a trade association. The statistical office publishes data for the whole industry in question, but the association produces tables for its members only. If all big enterprises but one belong to the association, it is easy to calculate values for the single enterprise that don't belong to the trade association. This kind of possibility is very difficult to prevent.

12.3   DISCLOSURE RISKS – MICRO DATA

Many statistical registers are very important for medical or social science research. Many researchers use registers where data from different fields are combined, and longitudinal registers. It is an important task for a statistical office to make such data available for research.

Registers used by research workers should be anonymised. When a register is released to researchers, it should be discussed which variables should be deleted. In registers on individual's name, address, personal identification number and real estate code should be taken out. In registers on enterprises, industry sector, size category and region can also be sensitive variables that can make it possible to identify larger enterprises.

There are many ways to minimise the risk of disclosure, e.g. the following:

– Be restrictive with which variables should be included in the researchers' data.
– Group the spanning variables by as few categories as possible, i.e. instead of breakdown by parish, the data would only be broken down by counties.
– Allow researchers access to a sample from the current register population. This should always be the appropriate beginning for any cooperation. If the researcher can first get to know the data material, it can then be possible to create a larger data matrix with just the variables of interest.

However, the best way to protect the confidentiality of microdata is to use a combination of remote access, legislation and licensing agreements.

Remote access means that researchers will only be able to analyse anonymised microdata at their own workplaces with access via the Internet. In this case, the microdata always remain at the statistical office. The researchers will not have access to microdata but can apply statistical techniques to the data and will then only get the results of the analysis done.

Those who get permission to analyse these microdata should have at least the same legal obligations as the staff at the statistical office to protect individual persons and enterprises, to ensure that sensitive information is not disclosed. The researchers will get the permission only after a legally binding agreement.

CHAPTER 13

Coordination and Coherence

It is important for register-based surveys that the statistical office has an organisation that supports extensive cooperation between those responsible for the register system's different parts. Managers play an important role in this coordination, among other things to ensure that the different parts of the register system can work together. However, an appropriate formal organisation is not sufficient. The organisational culture also needs to be one in which cooperation over organisational boundaries is seen as inherent.

Those working with register-based statistics must have a common approach and should not think only in terms of their own register. They must also be able to understand their role within the system. There is on one hand the responsibility to support the work of other registers, and on the other the possibility to use the system to develop new applications.

To develop such understanding, it is important to provide training, joint seminars, joint projects and tasks. Internal mobility within the organisation should also be encouraged. It is an advantage if there is a large number of people with experience of several registers within the system.

The cooperation mentioned above is an important condition when developing coordinated registers and surveys. The purpose of coordination is coherent estimates. With coordinated registers and suitable estimation methods, coherent estimates can be produced. This coordination and system-wide estimation methods are discussed in this chapter.

13.1 CONTENT-RELATED COORDINATION

To ensure that statistical data from different surveys are comparable, it is necessary that the populations and variables in the different surveys are *defined* in a consistent way. Furthermore, the registers should be *processed* with the aim of creating consistency between register populations and variables in different statistical registers. The coordination of content is based on the following:

– standardised *populations*, and
– standardised *variables*.

13.1.1 Coordination and standardisation of populations

Chapter 5 describes the creation of register populations and the definition of objects. All populations should be created using the base registers. One important use of the base registers is the creation of *standardised populations*, which are then used to create register populations in the system's other registers.

These standardised populations should be of high quality regarding coverage and the important spanning variables. They are therefore created *after* the period or point in time that they refer to. As far as possible, all administrative information should be reported in and processed before the population is defined. The register containing the standardised populations should then not be changed. Then, if other registers exclusively use these standardised register populations, the register-based statistics will be consistent with regards to the object sets.

Several Swedish registers begin with the standardised population in the Population Register that refers to 31/12 every year. Section 5.6 describes how detailed and completely consistent regional register-based statistics can be produced.

Consistency is, however, more of a problem with economic statistics. Swedish enterprise surveys are, to a large extent, based on their own data collection using frame populations formed during the November prior to the survey period.

Sections 5.4.8 – 5.4.9 describe in general terms how problems with the frame population arise. The example below shows further how statistics regarding energy enterprises are affected by inconsistencies in the population.

Chart 13.1 Inconsistent frame populations

Income and expenditure for Swedish energy enterprises are surveyed, both within Business Statistics (BS) and Energy Statistics (EN). Consistency editing has shown that different identity numbers are included in the two frame populations. Some identity numbers are included in both, others in only one of the surveys.

Income 1998 as per :	BS	EN	Income 1999 as per :	BS	EN	Costs 1998 as per :	BS	EN	Costs 1999 as per :	BS	EN
Only BS	32%	0%	Only BS	16%	0%	Only BS	32%	0%	Only BS	18%	0%
Both	68%	76%	Both	84%	96%	Both	68%	84%	Both	82%	96%
Only EN	0%	24%	Only EN	0%	4%	Only EN	0%	16%	Only EN	0%	4%
Total	100%	100%	Total	100%	100%	Total	100%	100%	Total	100%	100%

Because the problems with the frame population are significant, it is considered desirable that a more current version of the Business Register is used for many surveys. As different surveys are based on the Business Register at different points in time, the populations of these surveys will also be different. The industry classification for the same enterprise can also be different, which means that the populations of energy enterprises can be different in different surveys.

13.1.2 Coordination and standardisation of variables

Within economic statistics, some inconsistencies are due to differences in variable definitions and measurement methods. The example in Chart 13.2 below gives a description of how the statistics on Swedish energy enterprises are affected by inconsistencies relating to variables. The measured values for the individual energy enterprise's income and expenditure can differ between the two surveys.

Chart 13.2 Inconsistent variables

Industry (see Section 9.2.2)	Income and expenditure	
Business Statistics (BS) reports according to the *principal* industry, i.e. it truncates the multi-valued industry variable. Energy income and energy expenditure can then be reported in the "wrong" industry	Enterprises that are included with the same identity number in both Business and Energy statistics:	
	Income 1999	**Costs 1999**
	as per BS: 108 billion	as per BS: 64 billion
Energy Statistics (EN) reports so that enterprises active in several industries only report energy expenditure and income to the energy industry	as per EN: 103 billion	as per EN: 51 billion
	Enterprise with greatest difference:	Enterprise with greatest difference:
	EN – BS = 2 billion	BS – EN = 6 billion

13.2   COHERENCE

The concept of *consistency* relates to agreement. Two surveys are consistent if they relate to the same population and if the objects have the same measurement values for common variables, with the same estimates.

The concept of *coherence* refers to the fact that estimates from different surveys can be used together. For example, for a ratio to be meaningful, the numerator and the denominator must be coherent. Consistent surveys give coherent estimates.

Coherence is included as one of the quality components by many statistical offices. Through coordination and suitable estimation methods, coherence can be improved.

The statistical office in Holland has developed a new approach prioritising the consistency of the statistics. This is described in Statistics Netherlands (2000). Statistics from different sources can be consistent, i.e. have a high level of coherence, according to the Dutch methodology through:

– ensuring consistency regarding *populations* (relating to definitions of both object and object set),

– ensuring consistency regarding *variables*, and

– using calibration methods that give consistent *estimates*.

To ensure that the statistics are consistent, the discussion should not be limited to register-based statistics, but should include all types of surveys: sample surveys, censuses and register-based surveys should be coordinated.

Consistency regarding populations
To achieve consistency regarding populations, the object sets for all surveys referring to a specific object type and reference time should be coordinated. This can be done by creating a common register population for all these surveys. The common register population should have the highest level of quality possible, so it should only be created once all the relevant information has been received by the statistical office.

Consistency regarding variables
Definitions of variables should be coordinated. Related variables should be edited together, and the same corrections and imputations made in all surveys.

Calibration methods for consistent estimation
The method is described by Houbiers et al. (2003), and consists of repeated calibrations which give estimates that are consistent for all surveys concerned. The objects in the com-

mon population are divided into a series of blocks in which all the objects in each block have values for the same variables. Estimates are then formed for each block.

Beginning with a block with register information for all objects, estimates are then formed for the next best block, in terms of quality, that are consistent with the estimates in the register. The estimates for the second block are then made using the calibration method described in Section 7.5. The next step involves forming estimates for the third block that are consistent with the estimates from the preceding block, and so on.

13.3   CONSISTENT AND COHERENT ENTERPRISE STATISTICS

We use an enterprise example to illustrate a method for coordinated and consistent statistics. The aim is to show how inconsistencies can occur between surveys carried out at different points in time. If administrative data describing changes in the population are reported with a time delay, the frame errors will be large for the surveys carried out early. With the method presented here, it is possible to calculate revised estimates for all surveys based on a calendar year register.

All surveys in the example refer to year *t* and, for this year, a variety of surveys are carried out at different times. Firstly, quarterly surveys are carried out, based on the frame population created during November of year *t-1*. This survey refers to production in the manufacturing industry expressed as value added. Estimates are made using the variable *number of employees*, an auxiliary variable that exists in the sampling frame.

Then, at the beginning of year *t+1*, a census is carried out with its own data collection, based on the frame population created during November of year *t*. The value added of all enterprises is surveyed here. Finally, a register-based survey, based on data from the Statement of Earnings Register, is carried out. These data, referring to year *t*, become available during the spring of year *t+1*. The chart below shows the time when frames and register populations were created and the different surveys were carried out. In all surveys the object type is legal unit (LeU).

Chart 13.3 The chronological order of frames and surveys

November year *t-1*	A frame population (called frame I) is created with the current stock version of the Business Register containing all enterprises (legal units) known at that time
April year *t*	Sample survey using frame I of enterprises belonging to the manufacturing industry (D) regarding value added during 1st quarter year *t*
July year *t*	Sample survey using frame I of enterprises belonging to industry D regarding value added during 2nd quarter year *t*
October year *t*	Sample survey using frame I of enterprises belonging to industry D regarding value added during 3rd quarter year *t*
November year *t*	A frame population (called frame II) is created with the current stock version of the Business Register containing all enterprises (legal units) known at that time
January year *t+1*	Sample survey using frame I of enterprises belonging to industry D regarding value added during 4th quarter year *t*
January year *t+1*	Census using frame II of enterprises in all industries regarding value added during year *t*
Spring year *t+1*	Register-based survey for year *t* based on statements of earnings
Autumn year *t+1*	Calendar year population for year *t* is created with all legal units active during year *t* based on all available administrative information
Autumn year *t+1*	All surveys are revised with regard to the information in the calendar year register

13.3.1 Inconsistent populations

The Business Register receives information on enterprise restructuring, newly started and closed-down enterprises after a fairly long delay. Furthermore, data on the enterprise's industrial classification can also be incorrect in the November frame, which will then be detected later.

At some point during the autumn after the year in question, nearly all the information regarding the previous year has been received, and it is then possible to create a calendar year register of good quality. In the chart below, the frame errors of frame I and II are illustrated through a comparison with the final calendar year population for year t which is based on all information possible during the autumn year $t+1$.

Chart 13.4 Sampling frame I and II, calendar year version of the Business Register

Frame I: November year t-1			Frame II: November year t			Business Register's calendar year version. Calendar year t, created autumn *year $t+1$*			
Ent ID	Industry	No. empl	Ent ID	Industry	No. empl	Ent ID	Industry	No. empl	weight
LeU1	A	5	LeU1	A	5	LeU1	A	5	1
LeU2	D	210	LeU2	D	205	LeU2	D	205	1
LeU3	D	40	LeU3	D	45	LeU3	D	45	1
LeU4	D	120	LeU4	D	110	LeU4	D	110	1
LeU5	D	15				LeU5	D	15	0.25
LeU6	D	30	LeU6	D	25	LeU6	D	25	1
LeU7	E	55	LeU7	E	60	LeU7	E	60	1
LeU8	E	70	LeU8	E	65	LeU8	E	65	1
LeU9	F	90	LeU9	F	80	LeU9	F	80	1
LeU10	F	5	LeU10	F	10	LeU10	F	10	1
LeU11	G	340	LeU11	G	330	LeU11	G	330	1
LeU12	G	15	LeU12	G	20	LeU12	G	20	1
LeU13	G	10	LeU13	G	5	LeU13	G	5	1
LeU14	H	60	LeU14	H	70	**LeU14**	**K**	70	1
LeU15	K	20	LeU15	K	15	LeU15	K	15	1
LeU16	K	5	LeU16	K	10	LeU16	K	10	1
			LeU17	**D**	**20**	LeU17	D	20	0.5
						LeU18	**G**	**10**	**0.5**

Frame I consists of the current stock of active legal units during November year t-1 according to the information available at that moment.

Frame II consists of the current stock of active legal units during November year t according to the information available at that moment. The legal unit with identity 5 is not active during November and a new legal unit with identity 17 has been added.

The calendar year register consists of all legal units active during some part of year t. It is created when all administrative information regarding year t is available during the autumn of year $t+1$. Legal unit 5 has been active during the 1st quarter of year t, legal units 17 and 18 have been active during the last two quarters of year t and it is discovered that legal unit 14 does not belong to industry H, but to industry K. The time a legal unit is active during year t generate the weights in the calendar year register in the same way as in Chart 9.2.

The frame errors of frames I and II are large, which corresponds to our experiences of the Swedish Business Register (see Chart 1.8). With five sources we created a calendar year version of the Business Register for Sweden containing all enterprises (legal units) active during 2002. Each source consists of the legal units in one taxation system. In the table below, undercoverage and overcoverage of the sources are compared with our final calendar year version of the Business Register.

Only source 1 was available during November 2001 and was used as frame for sample surveys during 2002. The administrative object sets in each source is adequate for each of the five taxation systems. But taken alone, each source is of low *statistical* quality; however, if all sources are combined into a calendar year register, the coverage is good.

Chart 13.5 Overcoverage and undercoverage in five administrative sources

	Source 1	Source 2	Source 3	Source 4	Source 5
Overcoverage	41%	0%	0%	0%	0%
Undercoverage	21%	74%	74%	30%	9%

13.3.2 Preliminary estimates

The procedure is as follows:

– With sample surveys, as shown in Chart 13.6, preliminary estimates relating to the *industry D's* value added during year t are made. With equal probability, three enterprises are chosen from a total of five. Weights are calibrated using *number of employees* as an auxiliary variable. The weights w_i are given by calibrating d_i with *number of employees*, as described in Section 7.5. The enterprise with the identity *LeU5* is treated as nonresponse.

– With the census, shown in Chart 13.7, preliminary estimates regarding value added during year t are made.

Chart 13.6 Sampling frame I and sample survey regarding industry D, year t

A. Frame I: November year *t*

Ent ID	Industry	No. empl
LeU1	A	5
LeU2	D	210
LeU3	D	40
LeU4	D	120
LeU5	D	15
LeU6	D	30
LeU7	E	55
LeU8	E	70
LeU9	F	90
LeU10	F	5
LeU11	G	340
LeU12	G	15
LeU13	G	10
LeU14	H	60
LeU15	K	20
LeU16	K	5

B. Quarterly sample survey, value added industry D, year *t*

Value added, SEK m	Q 1	Q 2	Q 3	Q 4
LeU2	25	25	22	25
LeU4	15	12	7	14
LeU5	2	no response	no response	no response

C. Weights for the different estimates

	d_i	Q 1: w_i	Q 2–4: w_i
LeU2	5/3	1.09	1.18
LeU4	5/3	1.34	1.39
LeU5	5/3	1.63	0.00

D. Estimated value added

Q 1	50.7
Q 2	46.2
Q 3	35.7
Q 4	49.0
Total year 2	**181.7**

Two enterprises result in nonresponse in the census in Chart 13.7 below. The values for these are imputed using the ratio *Value added/Employee,* which is calculated using the enterprises that have responded from the same industry. The imputed value for *LeU10* is calculated as 10 · 26 / 80 = SEK 3.25 million.

Chart 13.7 Frame population II and census, year t

E. Frame II: November year *t*			F. Census, year *t* Value added, SEK million		G. Estimates for year *t*		
Ent ID	Industry	No. empl	Before imputation	After imputation	Industry	No. empl	Value added
					A	5	2
LeU1	A	5	2	2	D	405	**192**
LeU2	D	205	97	97	E	125	193
LeU3	D	45	21	21	F	90	29.25
LeU4	D	110	52	52	G	355	121
					H	70	60
LeU6	D	25	12	12	K	25	21.67
LeU7	E	60	93	93	Total	1 075	618.92
LeU8	E	65	100	100			
LeU9	F	80	26	26			
LeU10	F	10	no response	**3.25**			
LeU11	G	330	112	112			
LeU12	G	20	7	7			
LeU13	G	5	2	2			
LeU14	H	70	60	60			
LeU15	K	15	13	13			
LeU16	K	10	no response	**8.67**			
LeU17	D	20	10	10			

The estimates in table G above have been made by summing the values after imputation for each industry in the data matrices E and F.

The estimates for industry D are inconsistent, as the sample survey gives the annual total as SEK 181.7 million while the census gives SEK 192 million.

13.3.3 Revised and consistent estimates

During the autumn of year *t+1*, a standardised enterprise population is created, including all enterprises that have been active at any time during year *t*. For enterprises in this calendar year population, microdata are compiled from all four surveys in the same data matrix. This is shown in Charts 13.8 and 13.9 below. Finally, consistent estimates are made for year *t*; these estimates are shown in Chart 13.10.

The data matrix in Chart 13.8 below is therefore completely *consistent with regard to population.* The enterprise population has changed: firstly, the enterprise *LeU18* that was set up during the year has been added, and the industry code for enterprise *LeU14* has changed from H to K. The data matrix in Chart 13.8 is therefore *consistent with regard to the spanning variable industry.* The preliminary and incorrect industrial codes in Charts 13.6 and 13.7 have been corrected for all surveys.

In addition, the calendar year register contains a variable *weight,* which shows for how long the enterprise has been active during the year. The data matrix also contains wage sums based on administrative data from the Statement of Earnings Register. This register-based survey does not need any revision in this example, and was therefore not mentioned in Section 13.3.2 above.

Certain problems remain in the data matrix in Chart 13.8:

– The measured values relating to the value added for enterprise *LeU4* are different in the sample survey and the census.

– There are two imputed values in the census for *LeU10* and *LeU16*; the imputation model can be improved as the administrative source with wage sums now is available.

– Due to frame errors two values (*LeU5* and *LeU18*) are missing in the census, which is based on the frame from November of year *t*.

Chart 13.8 Calendar year population and data matrix with all surveys for year t

Business Registers' calendar year population				Sample, value added yr *t*, SEK m					Census yr *t*	Adm. data
Ent ID	Industry	No. empl	weight time of year	Q1	Q2	Q3	Q4	Year *t*	Value added SEK m	Wage sum SEK m
LeU1	A	5	1						2	0.3
LeU2	D	205	1	25	25	22	25	97	97	41.8
LeU3	D	45	1						21	9.2
LeU4	D	110	1	15	12	7	14	48	52	22.4
LeU5	D	15	0.25	2	missing	missing	missing	missing	missing	0.9
LeU6	D	25	1						12	5.1
LeU7	E	60	1						93	13.1
LeU8	E	65	1						100	14.2
LeU9	F	80	1						26	14.5
LeU10	F	10	1						(3.25)	1.8
LeU11	G	330	1						112	51.8
LeU12	G	20	1						7	3.1
LeU13	G	5	1						2	0.8
LeU14	K	70	1						60	13.5
LeU15	K	15	1						13	2.9
LeU16	K	10	1						(8.61)	1.9
LeU17	D	20	0.5						10	2.0
LeU18	G	10	0.5						missing	0.9

These problems are solved in the following ways:

LeU4:

There is an inconsistency in Chart 13.8 regarding the variable value added, because the sample survey gave SEK 48 million as the annual value for *LeU4*, while the census gave SEK 52 million. We assume that the census is more reliable and therefore change the quarterly values for *LeU4*. Each quarterly value is multiplied by 52/48.

LeU5:

The value 0 is given for *LeU5* for quarters 2–4 and value 2 for the full year.

LeU10, LeU16 and LeU18:

The values that are missing in the census are replaced with imputed values based on the ratio: *Value added/Wage sum*, which is calculated using the enterprises in the same industry that have responded.

The imputed value for *LeU10* is calculated as $1.8 \cdot 26 / 14.5 = $ SEK 3.23 million.

The imputed value for *LeU16* is calculated as $1.9 \cdot (60+13) / (13.5+2.9) = 8.45$ million.

The imputed value for *LeU18* is calculated as $0.9 \cdot (112+7+9) / (51.8+3.1+0.8) = 1.96$ m.

Once these changes have been made, the data matrix is complete. The final data matrix is shown in Chart 13.9 below.

Chart 13.9 Completed data matrix with all surveys for year t

| Business Registers' calendar year population | | | | Sample, value added yr *t*, SEK m | | | | | Census yr *t* | Adm. data |
Ent ID	Industry	No. empl	weight time of year	Q1	Q2	Q3	Q4	Year *t*	Value added SEK m	Wage sum SEK m
LeU1	A	5	1						2	0.3
LeU2	D	205	1	25	25	22	25	97	97	41.8
LeU3	D	45	1						21	9.2
LeU4	D	110	1	16.25	13.00	7.58	15.17	52	52	22.4
LeU5	D	15	0.25	2	0	0	0	2	2	0.9
LeU6	D	25	1						12	5.1
LeU7	E	60	1						93	13.1
LeU8	E	65	1						100	14.2
LeU9	F	80	1						26	14.5
LeU10	F	10	1						**3.23**	1.8
LeU11	G	330	1						112	51.8
LeU12	G	20	1						7	3.1
LeU13	G	5	1						2	0.8
LeU14	K	70	1						60	13.5
LeU15	K	15	1						13	2.9
LeU16	K	10	1						**8.45**	1.9
LeU17	D	20	0.5	Undercoverage in sample survey					10	2.0
LeU18	G	10	0.5						**1.96**	0.9

The data matrix in Chart 13.9 above is consistent with regard to *population* and *variables*.

The next step is to make consistent *estimates* for all surveys. The number of full-year employees is estimated using weights that show how much of the year the enterprise was active (see Section 9.1). Wage sums and value added are estimated by forming totals for each industry. The quarterly sample surveys are recalibrated (the calculations are not shown here) so that the estimates agree with the annual survey.

Finally, ratios can be formed, i.e. productivity measured as value added per full-year employee. This ratio uses numerators and denominators from different surveys. It is therefore a great advantage if the surveys are consistent on a micro level in the above data matrix.

Chart 13.10 Consistent estimates based on the consistent data matrix

Industry	Number of full-year employees	Wage sums SEK million	Value added SEK million	Value added per employee SEK million	Annual salary per employee SEK 000s
A	5.0	0.3	2.0	0.40	60.0
D	398.8	81.4	194.0	0.49	204.1
E	125.0	27.3	193.0	1.54	218.4
F	90.0	16.3	29.2	0.32	181.1
G	360.0	56.6	123.0	0.34	157.2
K	95.0	18.3	81.5	0.86	192.6
Total	**1 073.8**	**200.2**	**622.6**	**0.58**	**186.4**

This example shows the principles for how to create consistent enterprise statistics. However, this is just the first step towards consistent and coherent statistics. The number of full-year employees in the Business Register must also be consistent with the Employment Register and the wage sums need be consistent with the Statement of Earnings and Income Registers.

CHAPTER 14

Conclusions

The previous chapters in this book contain many proposals for change. New terms and new methods have been presented with the aim that register systems and register-based statistics can be developed and function in a better way than they do today. Dillman (1996) is rather pessimistic regarding innovation and change in government survey organisations, especially when dealing with non-sampling issues. One reason for this is the gap between operations and research cultures. We agree with Dillman that it is difficult to bring about change.

The implementation of new methods must be supported not only by managers, but also through a dialogue between the researchers developing new methods, and those working with the surveys that should be improved. The methods we propose have been developed at the same time as we have discussed register issues with those operating register products. We have spent much time in seminars and study circles to promote new ideas and methods. This implementation work will be necessary in all statistical offices, where new register-based statistics will be developed.

A new approach is necessary!

A new approach towards administrative data is necessary:

– There should be no prejudice that administrative data are of bad quality. If we compare the quality of the huge amounts of administrative data that tax authorities collect via tax forms from individuals and enterprises with the quality of the same data collected by the statistical office, we must admit that the tax authorities collect the best data. Scheuren and Petska (1993) are of the opinion that 'the detailed income and expenditure data on tax returns are generally regarded as more reliable than similar survey data'.

– On the other hand, administrative data should not be used as they are, they should be processed so that they can be used for statistical purposes. The most important part of this processing is the integration of many sources.

A new approach towards registers and statistical science is necessary. Some statisticians say there are no special methodological issues related to register statistics, and that there is no difference compared with ordinary censuses. This is a misunderstanding caused by the fact that they have not noticed the methodological issues which are unique to register-based surveys. The integration phase in register-based surveys and the methods used here to a large extent determine the quality and have no similarity with the methods used for censuses.

The development of register-based statistics should therefore be recognised as an important field for statistical science.

The register system

Administrative data from many sources are used to create a system of coordinated statistical registers. This register system can then be used to produce register-based statistics, create new registers and create frames for sample surveys or censuses. If the register system has been created in the right way, it will be an important factor promoting consistence and coherence between all surveys conducted at a statistical office. Even countries which are new as a producer of register-based statistics will benefit from coordinating the registers they have into one system.

It should be noted that we propose that *one* system is created for all surveys. If different subsystems are created (e.g. one system for social statistics and one system for economic statistics), then it will be difficult to combine data from these two systems. Productivity by industry can be estimated by dividing:

– value added by industry, from a register-based survey based on the Business Register, with

– hours worked by industry, from the Labour Force Surveys which is a sample survey based on the Population Register.

To achieve good quality here, the register system must be *one* coordinated system, and the estimation methods used must take into consideration that industry is a multi-valued variable. Integration errors, in this case aggregation errors, can give rise to inconsistencies between industry in the Business Register and industry in the Labour Force Survey.

Good coverage and consistency are the important advantages of register-based statistics if the register system has been created according to the principles in Chapters 5 and 6. All surveys based on the register system can benefit from this, and the National Accounts will get more consistent data without undercoverage.

Frame errors

Twelve out of thirteen chapters in Cochran (1963) are devoted to sampling errors. In the last chapter, Cochran mentions measurement errors and nonresponse. During the last few decades, there has been much effort spent on nonsampling errors; both measurement and nonresponse issues are today regarded as central issues in survey methodology.

In the book by Särndal and Lundström (2005), thirteen out of fourteen chapters are devoted to nonresponse issues, in the last chapter, frame errors are mentioned. Today, there are no established methods for handling frame errors; we believe that this kind of nonsampling error has been overlooked, and that the errors can be substantial. Development in this field is necessary, and these errors can only be reduced by register-statistical methods. If we learn how to create registers with good coverage, all surveys using these registers will benefit from the good coverage.

The first step is to become aware of the frame errors. At a statistical office, where sample survey theory is the dominating paradigm, registers are used to produce frames, and thereafter data are collected. As a rule, the quality of the frame population will never be known; instead new frames will be created followed by a new round of data collection.

A statistical office, where those responsible for, e.g., a business survey want to use administrative data, may follow exactly the same procedure, besides that, instead of sending questionnaires to the sampled enterprises, they use administrative data for the sample. In this way, response burden and costs will decrease, but the frame errors will be the same. If administrative data is used in this restricted way, the most important quality of administrative sources has not been used – the capacity for good coverage.

If registers are used both to create frames but also to create calendar year populations, then it will be possible to become aware of the frame errors. The preliminary estimates for the sample surveys based on frames can also be revised with information from the calendar year register, and the methods used to create frames can also be improved so that frame errors become smaller. The difference between a business frame based on early information and a calendar year business register based on all available administrative sources can be about 10% of total turnover, according to our experiences from the Swedish Business Register.

What more is needed?

Apart from developing the existing registers, staff at a statistical office should constantly be discovering *new administrative sources* that can be used to create new statistical registers and new products. New types of registers and databases that are created *outside the public sector* can also be relevant sources in the future. Individuals and enterprises will begin to leave numerous electronic tracks that are stored in databases by private enterprises. These types of register do not originate from relevant defined populations; the definition of an enterprise's database is determined by the enterprise's contacts with their customers, suppliers, etc. To investigate the position of an enterprise on the market, it is useful to be able to combine the data from the enterprise's own database with data referring to the relevant register population. This could potentially be an important role for national statistical offices in the future, *to be able to create relevant register populations* for different users' needs.

The ability to structure databases for statistical purposes and to analyse the data taken from administrative systems in a statistically meaningful way will be a skill required in many new fields. Register-statistical skills are therefore even required outside government survey organisations. *Universities and higher education* must pursue both research and carry out teaching regarding register-based surveys. This teaching and research should both relate to society and enterprise register-based statistics.

References

Bayard, K., Klimek, S. (2003): *Creating a Historical Bridge for Manufacturing Between the Standard Industrial Classification System and the North American Industry Classification System.* 2003 Proceedings of the American Statistical Association, Business and Economic statistics Section.

Bethlehem, J., Hoogland, J., Schouten, B. (2006): *From Surveys to Registers.* Paper presented for the fourth Seminar on Strategies for Social and Spatial Statistics, Oslo, Norway, 27-28 February 2006.

Biemer, P., Lyberg, L. (2003): *Introduction to Survey Quality.* John Wiley & Sons, Ltd.

Cochran, W. G. (1963): Sampling Techniques. 2nd edition, John Wiley & Sons, Ltd.

Deville, J., Särndal, C-E. (1992): *Calibration Estimators in Survey Sampling.* Journal of the American Statistical Association, 87, 376-382.

de Waal, T., Quere, R. (2003): *A Fast and Simple Algorithm for Automatic Editing of Mixed Data.* Journal of Official Statistics, Vol. 19, No. 4, 2003, pp. 383-404.

Dillman, D. (1996): *Why Innovation Is Difficult in Government Surveys.* Journal of Official Statistics, Vol. 12, No. 2, 1996, pp. 113-124.

Eltinge, J., Kozlow, R., Luery, D. (2003): *Imputation in Three Federal Statistical Agencies.* Paper prepared for the Federal Economic Statistics Advisory Committee, October 2003.

Eurostat (1997): *Proceedings of the Seminar on the Use of Administrative Sources for Statistical Purpose.* Luxembourg, 15-16 January 1997.

Federal Committee on Statistical Methodology (2005): *Report on statistical Disclosure Limitation Methodology.* Statistical Policy Working Paper 22, 2nd version.

Fellegi, I.P., Holt, D. (1976): *A Systematic Approach to Automatic Edit and Imputation.* Journal of the American Statistical Association, 71, 17-35.

Granquist, L., Kovar, J. (1997): *Editing of Survey Data: How Much Is Enough?* Lyberg et al. (eds.), Survey Measurement and Process Quality, John Wiley & Sons, Ltd.

Greijer, Å. (1995): *Overcoverage of foreign born in the Population Register estimated with the Labour Force Survey* (in Swedish). Metodrapport från BoR-avdelningen 1995:3, Statistics Sweden.

Greijer, Å. (1996): *Overcoverage in the Population Register – an analysis of returned mail* (in Swedish). Metodrapport från BoR-avdelningen 1996:7, Statistics Sweden.

Greijer, Å. (1997a): *Overcoverage in the Population Register estimated with the Income register* (in Swedish). Metodrapport från BoR-avdelningen 1997:11, Statistics Sweden.

Greijer, Å. (1997b): *Overcoverage in the Population Register estimated with the Labour Force Survey* (in Swedish). Metodrapport från BoR-avdelningen 1997:12, Statistics Sweden.

Groves, R., Fowler, F., Couper, M., Lepkowski, J., Singer, E., Tourangeau, R. (2004): *Survey Methodology*. John Wiley & Sons, Ltd.

Heady, P., Clarke, P., Brown, G., Ellis, K., Heasman, D., Henell, S., Longhurst, J., Mitchell, B. (2003): *Small Area Estimation – Project Report.* Model-Based Small Area Estimation Series No 2, Office for National Statistics.

Holt, D. (2001): *Comment to Platek and Särndal.* Journal of Official Statistics, Vol. 17, No. 1, 2001, pp. 55-61.

Houbiers, M., Knottnerus, P., Kroese, A.H., Renssen, R.H., Snijders, V. (2003): *Estimating consistent table sets: position paper on repeated weighting.* Statistics Netherlands, Discussion paper 03005, 2003.

Hörngren, J. (1992): *The Use of Registers as Auxiliary Information in the Swedish Labour Force Survey.* R&D Report, Statistics Sweden 1992:13.

Johansson, D. (1997): *The Number and the Size Distribution of Firms in Sweden and Other European Countries.* IUI Working Paper no 483, Research Institute of Industrial Economics, Stockholm Sweden, 1997.

Johansson, D. (2001): *The Dynamics of Firm and Industry Growth – The Swedish Computing and Communications Industry.* Royal Institute of Technology Stockholm Sweden, TRITA-IEO R 2001:05.

Kardaun, J. W. P. F., Loeve, J. A. (2005): Longitudinal *analysis in statistical offices.* Statistics Netherlands, discussion paper 05010.

Laan, P. van der (2002): *Creating a Social Statistics Database in the Netherlands: Progress and Priorities.* Paper presented for the second Seminar on Strategies for Social and Spatial Statistics, Copenhagen, Denmark, 5 September 2002.

Lindström, H. L. (1999): *Quality assurance in registers based on administrative sources for statistical purposes* (in Swedish). Rapport från Registerprojektet, Statistics Sweden.

Nanopoulos, P. (2001): *Comment to Platek and Särndal.* Journal of Official Statistics, Vol. 17, No. 1, 2001, pp. 77-86.

Pannekoek, J., de Waal, T. (2005): *Automatic Edit and Imputation for Business Surveys: The Dutch Contribution to the EUREDIT Project.* Journal of Official Statistics, Vol. 21, No. 2, 2005, pp. 257-286.

Platek, R., Särndal, C-E. (2001): *Can a Statistician Deliver?* Journal of Official Statistics, Vol. 17, No. 1, 2001, pp. 1-20.

Statistics Sweden (2001): *The future development of the Swedish register system – Final report of the Register Project and decision of the Director-General.* R&D Report, Statistics Sweden 2001:1.

Statistics Sweden (2001a): *Quality Concepts for Official Statistics.* MIS 2001:1, Statistics Sweden.

Scheuren, F. (1999): *Administrative Records and Census Taking.* Survey Methodology, Vol. 25, No. 2, pp. 151-160

Scheuren, F., Petska, T. (1993): *Turning Administrative Systems into Information Systems.* Journal of Official Statistics, Vol. 9 No. 1, pp. 109-119.

Selander, R., Svensson, J., Wallgren, A., Wallgren, B. (1998): *Administrative Registers in an Efficient Statistical system – New Possibilities for Agricultural Statistics? How Should We Use IACS Data?* Statistics Sweden and Eurostat 1998.

Statistics Canada (2003): *The Integrated Approach to Economic Surveys in Canada.* UN/ECE, Group of Experts on National Accounts, Eighth Meeting, April 2006.

Statistics Canada (2006): *Statistics Canada Quality Guidelines.* Fourth Edition – October 2003.

Statistics Denmark (1995): *Statistics on Persons in Denmark – A register-based statistical system.* Eurostat.

Statistics Finland (2004): *Use of Registers and Administrative Data Sources for Statistical Purposes – Best practices of Statistics Finland.*

Statistics Netherlands (2000): *Special issue – Integrating administrative registers and household surveys.* Netherlands Official Statistics, vol 15.

Statistics Netherlands (2004): *The Dutch Virtual Census of 2001 – Analysis and Methodology.*

Särndal, C-E., Lundström, S. (2005): *Estimation in Surveys with Nonresponse.* John Wiley & Sons, Ltd.

UN/ECE (1998): *Recommendations for the 2000 Censuses of Population and Housing in the ECE region.* United Nations 1998.

U.S. Census Bureau (2003): *U.S. Census Bureau Strategic Plan FY 2004-2008.* September 2003.

Wallgren, A., Wallgren, B., Perssson, R., Jorner, U., Haaland, J-A. (1996): *Graphing Statistics & Data – Creating Better Charts.* SAGE Publications 1996.

Wallgren, A., Wallgren, B. (1998*): Linking a system of time series – Adjusted estimates for the Labour Force Surveys 1987-1992* (in Swedish). Bakgrundsfakta till Arbetsmarknads- och Utbildningsstatistiken 1998:2, Statistics Sweden.

Wallgren, A., Wallgren, B. (1999): *Administrative Registers in an Efficient Statistical System – How Can we Use Multiple Administrative Sources?* Statistics Sweden and Eurostat.

Wallgren, A., Wallgren, B. (2002): *How to achieve good quality of register-based statistics?* Paper presented at the 2nd seminar for Social and Spatial Statistics, Copenhagen, Denmark, September 2002.

Wilén, A., Johannesson, I. (2002): *A new Total population register system – More possibilities and better quality.* Bakgrundsfakta till Befolknings- och Välfärdsstatistik 2002:2, Statistics Sweden.

Willenborg, L., de Waal, T. (2001): Elements *of Statistical Disclosure Control.* Springer.

Glossary

In this book we have introduced a number of terms and principles with the aim to make the development of register-statistical methodology easier. New register-statistical terms have been introduced in almost every chapter. Our aim in this glossary is to give an overview of these terms.

Adjoined variable	A derived variable in a register created with variables from another register with *different* objects. The objects in the first target register can be linked to objects in the second source register in a *one-to-one* relationship or a *one-to-many* relationship.
Administrative register	Registers used for administrative purposes in an administrative information system. An administrative register should contain all objects to be administrated, the objects are identifiable and the variables in the register are used for administrative purposes.
Aggregated data	Relates to summarised data, macrodata, for groups of objects and is usually presented as the content in statistical tables.
Aggregated variable	A derived variable in a register created with variables from another register with *different* objects. The objects in the source register can be linked to the objects in the target register using a *many-to-one* relationship. It is possible to aggregate values; in a way that is relevant for the survey, for the *micro objects* in the source register that is linked to the respective *macro object* in the target register.
Aggregation error	When we aggregate variable values from *many objects to one object*, this kind of error can occur if the variable is qualitative. Aggregation errors can arise from three reasons: objects occur several times in a register, many to one relations and multi-valued variables. Apart from causing erroneous estimates, aggregation errors also can give rise to inconsistencies between estimates from different registers.
Anonymised	By taking away all the identifying variables in a register, an anonymised data matrix is created.
Auxiliary variable	Register variable used during the estimation phase in a sample survey.
Base register	A statistical register of great importance for the whole register system. Base registers should define important object types, important object sets or standardised populations and contain links to objects in other base registers.

Register-based Statistics – Administrative Data for Statistical Purposes A. Wallgren and B. Wallgren

Calendar year register	The register version containing all objects that have existed at any point during a specific year. Objects that are added or cease to exist during the year should be included with information on the date of the event.
Calibration	An estimation method using weights, where the weights are adjusted to minimise the effects of various errors.
Census	Survey with own data collection, in which measurement values for all objects in the population are collected. Census-based registers are included in the register system.
Classification	Branch of industry, product category, education, occupation, etc. are examples of important statistical standards and classifications. These are based on international recommendations, are important in terms of content and are used in many surveys, both register-based studies and others. The administrative sources contain data on these classifications and this information is used to create variables within the register system.
Classification database	Formalised metadata to manage all the codes for the important statistical standards/classifications and keys between the different versions of these standards.
Classification error	The estimated *gross classification error* in is an estimate of the share of incorrect classifications in the entire register, where the estimate of the *net classification error* is an estimate of the systematic error.
Coding	Data in text form can be used to create statistically useful variables. The information is transformed from unstructured text to structured variable values in a coding process.
Coherence	Coherence refers to that estimates from different surveys can be used together. E.g., for a ratio to be meaningful, the numerator and the denominator must be coherent.
Cohort	A group of objects that originate from a specific period. Longitudinal registers can be analysed by following different cohorts over time.
Combination object	Estimation method for multi-valued variables in which every combination of objects and values of the multi-valued variable correspond to a combination object in a new data matrix where weights are used for estimation.
Communication variable	Communication variables such as name, address and telephone number are used when a statistical office needs to contact an object regarding a questionnaire or an interview.
Conceptual model	A conceptual model describes the system's object types and relations independently of physical/technical implementations.
Consistency	Two surveys are consistent if they relate to the same population and if the objects have the same measurement values for common variables, with the same estimates.
Consistency editing	In consistency editing of data in a register-based survey we are editing data from different sources, and suspected errors can both be caused by errors in variables and errors in objects.
Coverage error	Differences between the target population and the register population. See also overcoverage and undercoverage.
Cross-sectional quality	What comparisons can be made within the same register for a specific point in time/period?

Current stock register	Current stock register is the register version that is updated with all the available information related to currently active/living objects. Used as frame population.
Data matrix	Statistical data, microdata, sorted so that the matrix's columns contain observed variable values and the rows in the matrix contain observations for the objects.
Database	Databases that are used to store statistical microdata consist of a number of related database tables.
Database table	When Statistics Sweden migrated from mainframe computers to database servers, old terms such as flat file with records and positions were replaced by the term database table with rows and columns. A data matrix can be stored in a database table.
Definitions database	Formalised metadata to manage large quantities of variable definitions.
Demographic event	Objects are born, change location, alter or cease to exist. These types of events are called demographic events as they change the population.
Derived variable	New variables are formed using existing variables. Variables can be derived through grouping, classification, calculations, adjoining or aggregating. We also distinguish between local and imported derived variables
Derived object	Derived objects are created through register processing that uses existing information on relations. E.g.: Persons that are registered in the same dwelling form households.
Disclosure risk	If it is possible to derive sensitive characteristics about individuals or enterprises from a statistical table, the publication of these tables leads to risk for disclosure.
Documentation	The work to create metadata. See metadata.
Estimation	From an existing statistical register, tables with estimates can be produced and published. The term 'estimation' is used generally for sample surveys, but should also be used within register statistics. Here as well it is important to distinguish between the actual values and the estimates that the register produces.
	Within register statistics, the method of creating a register is the *fundamental* estimation method that determines which estimates can be done. To correct problems with multi-valued variables, overcoverage, missing values and time series level shifts, *supplemental* estimation methods can be used.
Events register	An event register for a specific period is a register containing information on all demographic events that have taken place during the period. A register is created for every type of event.
Events calendar	Changes or events that influence the register system should be documented. An events calendar is an IT system with formalised metadata in which it is possible to search for information on events related to time, register and variable.
Flow variable	Shows sums for different *periods* of time, e.g. salary income over one year.
Foreign key	Foreign keys are used to describe relations between different objects. See also reference variable.
Frame population	Refers to the object set defined by the frame and framing process. This term should only be used for censuses and sample surveys. The corresponding term for register-based surveys is register population.

Fundamental estimation methods	Fundamental estimation methods for register-based statistics consist of the methods for creating statistical registers – how to create the register population and the variables in the register.
Historical register	Registers containing information on all demographic events that have taken place for an object. An object that has three events, for example, can be found on three rows in the data matrix, etc.
Identifying variable	Identifying variables are used to clearly identify objects. The corresponding IT term is the primary key.
Imported variable	A variable taken from another statistical register.
Imputation	Missing values are replaced by imputed values that are calculated using a probability distribution or a deterministic model.
Imputation error	Imputed values that differ from the correct unknown value.
Integration	When a register is created, the work is dominated by the integration phase where data from many sources are combined. During this phase the register population and derived objects are created, variables are imported from different sources and derived variables are created.
Integrated data collection	An administrative authority collects data that is not used for administrative purposes, for a statistical authority. Statement of earnings data is an example of this, where the national Tax Board also collects certain information for Statistics Sweden.
Integrated register	Integrated registers are statistical registers that have been created by *only* combining information that already existed in the statistical registers in the system.
Integration error	The kinds of errors that have their origin from the integration phase should be called integration errors. In this category coverage errors, mismatch errors, missing values and aggregation errors are included.
Item nonresponse	Certain variable values can be missing for objects in the register. See missing values.
Link	Consists of one or several linkage variables that identify objects. If such variables are of good quality and exist in several registers, they can be used as links when several registers are matched. A link consists of information that identifies individual objects.
Linkage variable	A link between two registers consists of one or several linkage variables. A linkage variable consists of information that identifies a specific object or group of objects.
Linking of time series	An estimation method that aims to increase comparability over time.
Local primary variable	A primary variable that is directly taken from an administrative register is called a local primary variable.
Local variable	A variable that, for the first time in the register system, is formed within a certain register is a local variable for that register.
Locally derived variable	A derived variable that has been formed in the relevant register.
Longitudinal quality	What comparisons can be made at *micro level* over time?
Longitudinal register	An integrated register where registers for different points in time/periods have been compiled so that it is possible to follow the same object over time.

Macro metadata	Metadata for macrodata. Describes the content in statistical tables.
Macrodata	Refers to summarised data for groups of objects and is usually presented as the contents in statistical tables.
Matching	Links in two or more registers are compared. The result is a hit or non-hit (mismatch). Also called *exact* matching.
Matching error	Absent or false hits when matching.
Metadata	Information that is needed to be able to use and interpret statistics. Describes data by giving definitions of populations, objects, variables, the methodology and quality.
Micro metadata	Metadata for microdata. Describes the content in data matrices with microdata.
Microdata	Concerns data for individual objects.
Mismatch	See matching.
Missing values	Item nonresponse or missing values in registers can arise due to flaws in the administrative system, due to mismatch or due to that certain variable values are rejected during editing.
Model error	Derived variables can be created using a model. If the model's values differ from the correct values, model error occurs.
Multi-valued variable	A variable that has more than one value for at least one object. For example, the variable occupation can be described by one value for those who have one occupation but must be described with more than one value for those who have more than one occupation.
Natural random variation	Estimates based on registers can be affected by natural random variation, which is why differences and changes should not be interpreted in an uncritical way.
Nonresponse adjustment	Estimation method that aims to minimise nonresponse error or the effects of missing values.
Object	A population, register or data matrix consists of a number of objects (elements or statistical units).
Object instance	The corresponding term for *object* in conceptual modelling of an IT system is *object instance*. This terminology refers to a population of, e.g., individuals of object type person where every individual in the population is an object instance. The term object is often used as a synonym for object instance.
Object set	When we describe a register without referring to a specific survey, we use the concept *object set*. However, every statistical register is created for one or several principle uses or surveys. It is therefore common that the register's object set agrees with the principle survey's population.
Object type	The type of object that a register or a population consists of. For example, person, household, local unit, property
Observation	All the measurement values for a specific object. An observation is also called record.
Overcoverage	The register contains objects that do not belong to the target population.

Population	The population definition should show which objects are included in that population. The object type should be specified. A time reference and geographic delimitation should always be included. The geographic delimitation should also specify the relation that exists between the object or statistical unit and the geographical area.
Primary key	IT term for identifying variable.
Primary variable	A variable can be a primary variable or a derived variable. Variables taken directly from an administrative register are called local primary variables.
Primary statistical register	Primary registers are statistical registers that are *directly* based on *at least one* administrative source. Primary registers are based on administrative sources and with these sources the main part of the statistical variables of the register system are created.
Reference variable	Reference variables (foreign keys) are used to describe relationships between different objects. When matching registers that contain data on different objects, reference variables produce hits between related objects.
Register	A register aims to be a complete list of the objects in a specific group of objects or population. However, data on some objects can be missing due to quality deficiencies. Data on an object's identity should be available so that the register can be updated and expanded with new variable values for each object. Complete listing and known identities are thus the important characteristics of a register.
– *Administrative register*	Registers used for administrative purposes in an administrative information system. An administrative register should contain all objects to be administrated, the objects are identifiable and the variables in the register are used for administrative purposes.
– *Base register*	A statistical register of great importance for the whole register system should define important object types, important object sets or standardised populations and contain links to objects in other base registers.
– *Calendar year register*	The register version containing all objects that have existed at any point during a specific year. Objects that are added or cease to exist during the year should be included with information on the date of the event.
– *Current stock register*	Current stock register is the register version that is updated with all the available information to relate to currently active/living objects.
– *End of year register*	A register version that refers to December 31 or January 1. Suitable for annual population statistics. An example of a register referring to a specific point in time.
– *Events register*	An event register for a specific period is a register containing information on all demographic events that have taken place during the period. A register is created for every type of occurrence.
– *Historical register*	Registers containing information on all demographic events that have taken place for an object. An object that has three events, for example, can be found on three rows in the data matrix etc.
– *Integrated register*	Integrated registers are statistical registers that have been created by *only* combining information that already existed in the statistical registers in the system.
– *Longitudinal register*	An integrated register where registers for different points in time/periods have been compiled so that it is possible to follow the same object over time.

– Primary statistical register	Primary registers are statistical registers that are directly based on at least one administrative source as. Primary registers are based on administrative sources and with these sources the main part of the statistical variables of the register system are created.
– Register at a specific point in time	The register referring to a specific point in time, such as the turn of the year, is the version of the register that is updated to describe the object set at that point in time. This update is carried out after the point in time, when information on all events up to that point in time is available. Is used for register-based surveys.
– Source register	The term source register refers to both the administrative sources and the Statistics Sweden registers that are used to create the new register.
– Statistical register	Registers that have been processed for statistical purposes. Statistical registers are created by processing administrative registers so that objects sets, objects and variables meet statistical needs.
Register maintenance survey	A questionnaire sent to the objects in a base register for which data is missing or can be considered out-of-date.
Register population	Refers to the object set in the register which has been created for the survey in question, i.e. the population being surveyed. The corresponding term for censuses and sample surveys is frame population
Register-based survey	A statistical survey without own data collection. Existing administrative or statistical registers are used instead.
Register-based statistics	Statistics produced using register-based surveys.
Relation	Between different objects of the same object type, there can be relations of different kinds, i.e. parent-child. There can also be relations between objects of different object types, i.e. employee-employer. When many relations are important from a legal point of view, data on relations are often included in the administrative register. With these relations, links in the register system can be created.
Relational object	A relation between two objects can be seen as a relational object. It is necessary to objectify a relation when there are variables that must be linked to combinations of objects rather than to one individual object.
Relevance error	If the definitions of a population and/or a variable are not adequate for the survey's aims, the surveys results are subject to relevance error.
Response variable	For every cell in a table descriptive measures are calculated for *response variables*.
Sample survey	A survey with own data collection, where measurement values for a probability sample of the population's objects are collected.
Selection	A new register is created by choosing certain objects and variables from existing registers. We differentiate between *object selection,* where certain objects are selected, and *variable selection,* where certain variables are selected.
Single-valued variable	A variable that can only have one value for every object instance.
Source register	The term source register refers to both the administrative sources and the statistical registers in the register system that are used to create a new register.

Spanning variable	Spanning variables define the cells in statistical tables. In a one-way table there is one spanning variable, in a two-way table there are two spanning variables etc.
Standard	See classification
Standardised population	All populations should be created using the base registers. One important use of the base registers is the creation of standardised populations, which are then used to create register populations in the system's other registers. Standardised populations should be of high quality regarding coverage and the important spanning variables. They are therefore created after the period or point in time that they refer to when all administrative has been reported. Then, if other registers exclusively use these standardised register populations, the register-based statistics will be consistent with regards to the object sets.
Standardised variable	Variables of key significance that are used within several registers. There is a clear responsibility for these variables with regards to definitions, naming and documentation.
Statistical register	Registers that have been processed for statistical purposes. Statistical registers are created by processing administrative registers so that objects sets, objects and variables meet statistical needs.
Stock variable	Shows characteristics at a specific point in time, i.e. age of a person at a specific point.
Supplementary estimation methods	Supplementary estimation methods for register-based statistics consist of estimation methods using a statistical register together with weights or imputations.
Table	(Statistical) tables contain macrodata and database tables contain microdata.
Technical variable	Variables for internal register administration.
Time reference	Variable showing a point in time for an event that affects objects or updates in the register.
Time series quality	Which comparisons can be made over time on an aggregated level?
Undercoverage	Objects that belong to the target population are missing in the register.
Value set	The set of values that a variable can take on, or can be thought to take on, for any object.
Variable	A variable is a measurable attribute of an object.
– *Adjoined variable*	A derived variable in a register created with variables from another register with *different* objects. The objects in the first target register can be linked to objects in the second source register in a *one-to-one* relationship or a *one-to-many* relationship.
– *Aggregated variable*	A derived variable in a register created with variables from another register with *different* objects. The objects in the source register can be linked to the objects in the target register using a *many-to-one* relationship. It is possible to aggregate values, in a way that is relevant for the survey, for the *micro objects* in the source register that is linked to the respective *macro object* in the target register.
– *Auxiliary variable*	Register variable used during the estimation phase in a sample survey.

– *Classification*	Branch of industry, occupation, etc. are examples of important statistical standards and classifications. These are based on international recommendations, are important in terms of content and are used in many surveys, both register-based studies and others. The administrative sources contain data on these classifications and this information is used to create variables within the register system.
– *Communication variable*	Communication variables such as name, address and telephone number are used when Statistics Sweden needs to contact an object regarding a questionnaire or an interview.
– *Derived variable*	New variables are formed using existing variables. Variables can be derived through grouping, classification, calculations, adjoining or aggregating.
– *Flow variable*	Shows sums for different periods of time, i.e. salary income over one year.
– *Identifying variable*	Identifying variables are used to clearly identify objects. The corresponding IT term is the primary key.
– *Imported variable*	A variable taken from another statistical register
– *Link*	Consists of one or several connecting variables that identify objects. If such variables are of good quality and exist in several registers, they can be used as links when several registers are matched. A link consists of information that identifies individual objects.
– *Linkage variable* 20–21, 24–25, 28,	A link between two registers consists of one or several linkage variables. A linkage variable consists of information that identifies a specific object or group of objects.
– *Local primary variable*	A primary variable that is directly taken from an administrative register is called a local primary variable.
– *Local variable*	A variable that, for the first time in the register system, is formed within a certain register is a local variable for that register.
– *Locally derived variable*	A derived variable that has been formed in the relevant register.
– *Multi-valued variable*	A variable that has more than one value for at least one object. For example, the variable occupation can be described by one value for those who have one occupation but must be described with more than one value for those who have more than one occupation.
– *Primary key*	IT term for identifying variable.
– *Primary variable*	A variable can be a primary variable or a derived variable. Variables taken directly from an administrative register are called local primary variables.
– *Reference variable*	Reference variables (foreign keys) are used to describe relationships between different objects. When matching registers that contain data on different objects, reference variables produce hits between related objects.
– *Response variable*	For every cell in a table descriptive measures are calculated for *response variables*.
– *Single-valued variable*	A variable that can only have one value for every object instance.
– *Spanning variable*	Spanning variables define the cells in statistical tables. In a one-way table there is one spanning variable, in a two-way table there are two spanning variables, etc.
– *Standardised variable*	Variables of key significance that are used within several registers. There is a clear responsibility for these variables with regards to definitions, naming and documentation.

– *Stock variable*	Shows characteristics at a specific point in time, i.e. age of a person at a specific point.
– *Technical variable*	Variables for internal register administration.
– *Time reference*	Variable showing a point in time for an event that affects objects or updates in the register.
– *Weight-generating variable*	Register variable which is used to create the weights needed to make estimates for multi-valued variables.
Variable selection	See selection
View	Database term for an adjusted presentation of the content in database table(s).
Weight	In data matrices for samples, every observation in the sample can represent many observations in the population. An important part of the work with sample surveys is to calculate these weights or adjustment coefficients.
Weight-generating variable	Different register variables can be used to create the weights needed to make estimations for polyvalent variables.

Index

Printed and bound in the UK by
CPI Antony Rowe, Eastbourne

Printed and bound by CPI Group (UK) Ltd, Croydon, CR0 4YY

16/04/2025

14658550-0001